Minitab® Lab Manual

for Devore/Peck's

STATISTICS:
THE EXPLORATION AND
ANALYSIS OF DATA

FOURTH EDITION

Roger E. Davis
Pennsylvania College of Technology

DUXBURY

THOMSON LEARNING

Australia • Canada • Mexico • Singapore • Spain • United Kingdom • United States

Sponsoring Editor: *Carolyn Crockett*
Assistant Editor: *Seema Atwal*
Editorial Assistant: *Ann Day*
Production Editor: *Scott Brearton*
Marketing Team: *Samantha Cabaluna, Ericka Thompson*
Cover Design: *Vernon Boes*
Print Buyer: *Micky Lawler*
Cover Printing: *Webcom, Inc.*
Printing and Binding: *Webcom, Inc.*

For more information about this or any other Duxbury products, contact:
DUXBURY
511 Forest Lodge Road
Pacific Grove, CA 93950 USA
www.duxbury.com
1-800-423-0563 (Thomson Learning Academic Resource Center)

MINITAB is a registered trademark of Minitab, Inc.

For permission to use material from this work, contact us by
Web: www.thomsonrights.com
fax: 1-800-730-2215
phone: 1-800-730-2214

Printed in Canada

5   4   3   2   1

ISBN 0-534-38041-7

Table of Contents

Chapter 1
Getting Started With Minitab

1.1 Overview

This chapter covers the basic structure and commands of Minitab for Windows Release 11. After reading this chapter you should be able to

1. Start Minitab
2. Identify the Main Menu Bar
3. Enter Data into Minitab
4. Save the Data File
5. Compute Descriptive Statistics
6. Print the Session Window
7. Obtain Online Help
8. Exit Minitab.

Minitab commands and software features are featured in areas where they are appropriate for the specific statistical analysis.

1.2 Starting Minitab

Minitab is a computer software program initially designed as a system to help in the teaching of statistics, and over the years has evolved into an excellent system for data analysis. The procedure for starting Minitab requires only that you:

1. **Select Start>Programs>Minitab 13 for Windows>Minitab, or**

2. **Double click on the blue Minitab for Windows icon as shown in Figure 1.1.**

Figure 1.1

1.3 The Main Menu

The main Minitab window contains numerous subwindows, two of which are shown in Figure 1.2: the Worksheet window and the Session window. A third window is the Project Manager. The Project Manager contains folders that allow access to various parts of your project. These folders include Session, History, Graphs, Report Pad, Related Documents, and Worksheet folders. Across the top of the Minitab window is the menu bar, from which menus may be opened and from which you choose commands. The Session and Worksheet windows are the most important and the most frequently used windows.

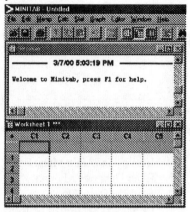

Figure 1.2

The main menu bar, shown in Figure 1.3, contains selections common to most Windows applications and some selections specific to Minitab. The File command contains options related to opening files, saving files, printing, and exiting Minitab.

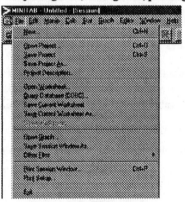

Figure 1.3

The Edit command contains options related to deleting, copying, and pasting. The other selections on the menu bar, Manip(ulate), Calc(ulate), Stat(istics), Graph are specific to Minitab. The final two selections on the main menu bar, Window and Help are found in most Windows applications. The Window command enables you to switch among windows, while the Help command enables you to get on-line help from Minitab.

1.4 Entering Data

Minitab's Worksheet window, as shown in Figure 1.4, is like a spreadsheet in that it works with data in rows and columns. Typically, a column contains the data for one variable, with each individual observation in a row. Columns are designated as C1, C2, C3,... and rows are numbered 1, 2, 3, ...

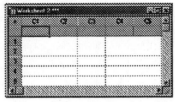

Figure 1.4

The size of the worksheet is limited only by the memory available and the size of the hard drive.

There are several ways to enter data into the Minitab Data window. You may **read data from a file** or **type in the data**. Let's look at a problem involving a combination of reading data from a file and adding data to the dataset by typing in the additional data.

The Problem

(Example 4.1 text) Traumatic knee dislocation often requires surgery to repair ruptured ligaments. One measure of recovery is range of motion (measured by the angle formed when, starting with the leg straight, the knee is bent as far as possible). The paper "Reconstruction of the Anterior and Posterior Cruciate Ligaments After Knee Dislocation" (*American Journal of Sports Medicine* (1999):189-197 reported on the post-surgical range of motion for a sample of 13 patients.

Reading Data from a File

Follow these steps to read data from a file:

1. Start Minitab
   Locate and double click on the Minitab program group icon and double click on the blue Minitab for Windows icon.

2. Open the file.
   Choose **File**>**Open Worksheet** from the menu. A portion of the submenu is shown in Figure 1.5.

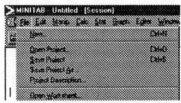

Figure 1.5

At the completion of this operation, all data in the current worksheet will be replaced with the data in the file. When you select **Open Worksheet** the dialog box shown in Figure 1.6 will open.

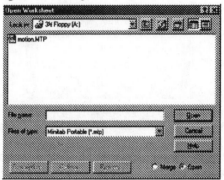

Figure 1.6

Minitab allows you to open files from many different software packages. Minitab worksheets use the file extension .mtw and Minitab portable files use the extension .mtp. Choose the location and the type of file you want to open, then select the filename from the list and choose **Open** to open the file. **Select the**

file from drive A:motion.mtp and choose **Open**.

Typing Data into the Worksheet
Follow these steps to add data to the dataset:

3. Make the Worksheet window the active window.
   Position the cursor in the Worksheet window in the column and cell where you want the data located. Position the cursor in column 1 row #.

4. Enter the data.
   Enter the data indicated below beginning in row 11 of column C1. Type in the data value and press {ENTER} after each entry.
   127   135   122

5. Correcting errors.
   If you enter an incorrect value, highlight the cell, retype the data entry and press {ENTER}. Do not delete the error, just type in the correct value. Deleting the data causes the entire column to move up one line!! Change the entry in column 1 row 12 from 135 to 134.

Naming Columns in the Worksheet
Columns are generally used for different variables within the dataset.
1. Name the column.
   To name a column (variable) in the worksheet, position the cursor in the box at the top of the column above row 1 and below the C# label. Type in the name you want to assign to the column. In version 13.1 of Minitab, column names may be longer than 8 characters. Position the cursor in the box at the top of column 1 above row 1 and type in the name Motion.

1.5   Compute Descriptive Statistics

Minitab offers a variety of basic statistics to analyze data. Let's begin by obtaining a summary table describing the variable Motion.

Follow these steps to determine the descriptive statistics for Height.
1. Compute descriptive statistics.

   Begin by using the mouse to click on **Stat>Basic Statistics>Display Descriptive Statistics**.

2. Enter the information in the dialog box, as shown in Figure 1.7.

   Select the variable Motion by highlighting Motion and double clicking (or Select). Choose **OK.**

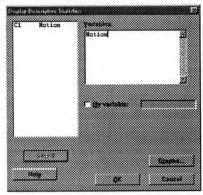

Figure 1.7

1.6   Saving a File

There are three basic components in a Minitab session: the worksheet (contained in the Worksheet window), the Session window and graphs. Saving graphs will be covered after you create your first graph.

Saving a Worksheet
Follow these steps to save a Minitab worksheet for the first time.
1. Choose **File>Save Worksheet As...**

2. Select the drive.

   Designate the correct drive and path for saving the file in the dialog box, as shown in Figure 1.8, then position the cursor in the box labeled File Name:

5

and type in the filename.

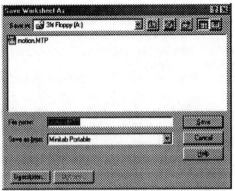

Figure 1.8

Minitab uses the same file naming conventions as Windows. Minitab work-sheets use the file extension .mtw and Minitab portable files use the extension .mtp. Choose **Save** to save the file. Designate the drive as A: and enter the filename motion. Choose **Save.**

Saving a Project

When you save your work as a project, you save all the information about your work. The contents of every window is saved, including the columns of data in each Worksheet window, the complete text in the Session window and History window, and each Graph window. You will want to save these results if it is necessary to examine the output at a later time or use the output in a document. **Follow these steps** to save a project.

a. Choose **File>Save Project As...** Designate the correct drive and path for saving the file in the dialog box, as shown in Figure 1.9, then position the cursor in the box labeled File **N**ame: and type in the filename for this project file. Select drive A: and enter the filename motion.mpj and choose

Save.

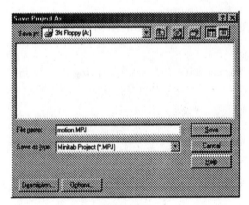

Figure 1.9

1.7   Printing the Session Window

Follow these steps to print a copy of the Session window.

1. Make the Session window the active window.
   Click on the title bar of the Session window to make it the active window.

2. Select the correct printer.
   If necessary, choose **File**>**Print Setup...**   to select the correct printer. After selecting the correct printer chose **OK.**

3. Print the Session window.
   Choose **File**>**Print Session Window.**.(see Figure 1.10).   Chose **OK** to print the Session window.

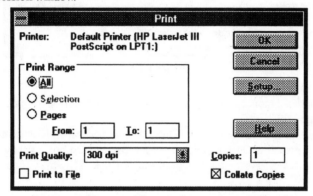

Figure 1.10

1.8   Obtaining On-line Help

Follow these steps to obtain on-line help.

1. Click with the mouse on **Help**>**Help**, to bring up the dialog box shown in Figure 1.11.

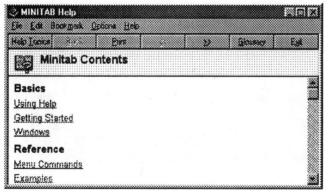

Figure 1.11

2. Select the topic.

7

Double click on the text "Getting Started" and "Introduction to Minitab". The Help window, as shown in Figure 1.12, displays basic information on using Minitab. Click on the button labeled "Exit" near the top of the Help window.

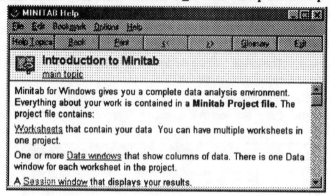

Figure 1.12

3. Using Search.

Click with the mouse on **Help**>**Search Help**, to bring up the dialog box shown in Figure 1.13.

Type hist in the first text area. Click the line that says "Histogram (Graph menu)." From the Click a topic, then click Display text box, as shown in Figure 1.14, select "Histogram (Graph menu)." Click "Display." The results are shown in Figure 1.15.

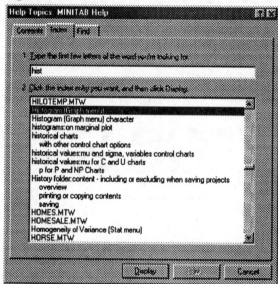

Figure 1.13

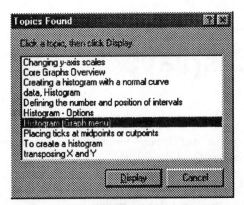

Figure 1.14

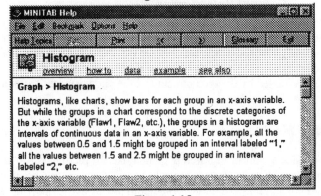

Figure 1.15

4. Exiting Help.

To return to the Minitab session, choose the **Ex**it button from the menu bar or choose **Fi**le>**Ex**it.

Chapter 2
Data Displays

2.1 Overview

This chapter covers the basic displays for categorical and numerical data. After reading this chapter you should be able to

1. Construct a Histogram (Bar Chart) for Categorical Data
2. Construct a Stem-and-Leaf Display
3. Construct Frequency Distributions
4. Construct Histograms

One of the most useful ways to begin an initial exploration of data is to use techniques that result in a pictorial representation of the data. The graphic representations can visually reveal characteristics of the variable being examined. There are a variety of graphic techniques that may be used to describe the data. The technique used is dictated by the type of data and the circumstances surrounding the problem.

2.2 Graphic Modes

Minitab provides a number of different graph types for plotting single and multiple variables, as well as statistical control charts. There are two graphic modes: low-resolution and high-resolution. While both modes may display the same information, the high-resolution mode offers pictorial elements like lines and colors and provides a professional graphic presentation. The high-resolution graph commands enable will enable you to create a virtually unlimited variety of graphs. Low resolution graphs are called character graphs are made up of normal keyboard characters. Low-resolution graphics can be viewed on any screen, printed on any printer, and stored in an outfile.

2.3 Stem-and-Leaf Displays

New Minitab Commands

1. **Graph>Stem-and-Leaf** - Produces a character-based stem-and-leaf plot in the Session window. In this section, you will construct stem-and-leaf plots for two data sets.

The stem-and-leaf display provides an opportunity to explore a data set containing numerical data that may be either discrete or continuous. This exploration provides you the opportunity to obtain an intuitive feel for the shape of the data. Such a preliminary organization often reveals useful information and opens up paths of inquiry. Let's look at the following problem to construct a basic stem-and-leaf plot.

The Problem - Paperback Mysteries

(Example 3.9) A sample of 40 paperback mysteries was randomly selected from the shelves of a local bookstore and the number of pages in each one was recorded. The resulting data is displayed in the order in which observations were obtained:

229	247	347	246	307	181	198	214	234	340
314	260	202	320	360	320	200	414	262	248
376	211	214	218	276	628	255	352	197	308
203	371	203	406	261	378	223	181	284	196

Follow these steps to construct a basic stem-and-leaf display for the Pages (using truncated leaves):

1. Enter data.

   Enter the sales amounts in column C1. Name column C1 as Pages.

2. Construct the stem-and-leaf display.

   Choose **Graph>Stem-and-Leaf**. Place Pages in the Variables: text box. Choose **OK**.

The Minitab Output

```
Stem-and-leaf of Pages N  = 40
Leaf Unit = 10


        5      1 88999
       19      2 00001111223444
      (6)      2 566678
       15      3 0012244
        8      3 56777
        3      4 01
        1      4
        1      5
        1      5
        1      6 2
```

Figure 2.1

The section of the session window labeled Stem-and-Leaf Display: Pages (Figure 2.1) indicates the number of observations (N) and depth information. To the left of the stem, Minitab indicates the cumulative number of observations, count-

11

ing in from the extremes. This is the depth information. The depth represents how far the observation on the right is from the appropriate end of the data set. For example, the value 240, is represented as the 4 on the 2 stem and is the seventeenth, eighteenth and nineteenth observations from the beginning of the ordered data set. The stem in which the median occurs is indicated in parentheses, and displays the frequency for that stem alone.

The stem-and-leaf display of Figure 2.1 indicates that the mean page length is around 280 pages. The data range from 181 to 628 pages with most of the data appearing between 200 and 300 pages. The data set appears to contain one outlier - 628 pages.

Crime Rates - a real data set

The file USCrime contains the cime index rates of violent crimes per 100,000 residents from 1960 to 1997. (Source: http://www.disastercenter.com/crime/uscrime.htm).

Follow these steps to construct a basic stem-and-leaf display for the variable Robbery:

1. Open the worksheet.

   Choose **File>Open Worksheet**. Select the file a:USCrime.mtp (included on the disk provided with the manual). Choose **Open**.

2. Construct the stem-and-leaf display.

   Choose **Graph>Stem-and-Leaf**. Place Robbery in the Variables: text box. Choose **OK**.

The Minitab Output

```
Stem-and-leaf of Robbery    N  = 38
Leaf Unit = 10


    2       0 55
    6       0 6667
    7       0 8
    8       1 0
    9       1 3
   10       1 4
   11       1 7
   18       1 8888999
   (7)      2 0000111
   13       2 2222333
    6       2 5555
    2       2 67
```

Figure 2.2

The objective of data displays is to engage in a conversation with the data, obtaining good information from the display about the data set. Typically, the data may be described numerically in measures of the center and measures of variability.

The stem-and-leaf display of Figure 2.2 indicates that the data is skewed somewhat to the right and there may be some outliers. The data set indicates that there are many rates of 180 robberies or greater per 100,000 inhabitants and a smaller number of rates less than 180 robberies per 100,000 inhabitants.

Exercises

2.1 Prices of concert tickets for a number of different concerts are given in the following table:

21	17	26	46	26	34
37	38	27	31	36	34
41	48	25	36	40	28
26	36	39	34	46	36
32	36	41	38	32	41

Construct a stem-and-leaf display of the data. Describe the shape of the data set: is the shape of the data symmetric; skewed right or left; are there multiple peaks, or outliers?

2.2 A random sample of mathematics placement test scores from a college are in the following table:

14	68	67	30	76	83	19	45	60	24
28	10	12	39	82	86	39	40	79	77
30	79	58	86	80	49	22	22	13	57
79	88	49	9	72	66	68	34	57	46
20	70	20	45	62	7	60	81	20	81

Construct a stem-and-leaf display of the data. Describe the shape of the data set: is the shape of the data symmetric; skewed right or left; are there multiple peaks, or outliers?

2.3 A department of transportation engineer has recorded the number of traffic accidents reported in the state for a number of days during January and February.

152	154	186	209	126	152	125	182	123	126
128	122	168	185	135	162	179	160	146	121
176	135	165	135	188	150	216	143	150	132
201	187	219	123	173	149	168	122	168	200
203	209	175	154	169	188	159	187	206	148

Construct a stem-and-leaf display of the data. Describe the shape of the data set: is the shape of the data symmetric; skewed right or left; are there multiple peaks, or outliers?

2.4 Light intensity near a highway sign at night influences our ability to read the sign. Measurements of light intensity (in candela per square meter) are contained in the file light.mtp. Construct a stem-and-leaf display of the intensity measurements. Describe the shape of the data set: is the shape of the data symmetric; skewed right or left; are there multiple peaks, or outliers?

2.4   Frequency Distributions

New Minitab Commands

1. **Stat>Tables>Tally** - Prints a one-way table of counts and percents for specified variables. Minitab displays summary information for each distinct value in the column. In this section, you will construct frequency distributions for categorical and numerical (discrete) data.

    Lists of large sets of data do not communicate much information. While stem-and-leaf displays are an appropriate exploratory data analysis tool, sometimes you will want to condense the data in the form of a frequency distribution. A frequency distribution is a table that pairs each category with its frequency and/or relative frequency.

Categorical Data

The Problem

(Example 3.1, text) Many public health efforts are directed toward increasing levels of physical activity. The paper "Physical Activity in Urban White, African-American, and Mexican-American Women" (*Medicine and Science in Sports and Exercise* (1997):1608-1614) reported on physical activity patterns in urban women. The accompanying data set gives the preferred leisure time activity for each of 30 Mexican-American women.

Walking	WeightTraining	Aerobics	Walking
Gardening	WeightTraining	Walking	Walking
Cycling	Walking	WeightTraining	Walking
Aerobics	WeightTraining	WeightTraining	Walking
Gardening	Walking	Walking	Cycling
Aerobics	Walking	Aerobics	Walking
Walking	Walking	WeightTraining	Walking
Walking	WeightTraining		

Follow these steps to construct a frequency table for the activities:

1. Open the worksheet.
   Choose **File>Open Worksheet**. Select the file A:Activity.mtp. Choose **Open**.

2. Construct the frequency table.
   Choose **Stat>Tables>Tally...** Place Activity in the **V**ariables: text box. Place a check in the **C**ounts checkbox and place a check in the **P**ercents checkbox. Choose **OK**.

The Minitab Output

Tally for Discrete Variables: Activity

```
  Activity  Count  Percent
  Aerobics      4    13.33
   Cycling      2     6.67
 Gardening      2     6.67
   Walking     15    50.00
WeightTraining  7    23.33
        N=     30
```

Figure 2.3

The section of the session window, as shown in Figure 2.3, labeled Tally for Discrete Variables indicates the type of activity, the count and the percent in each category.

Numerical Data

Discrete data - The Problem

Scores for the first exam in two sections of a statistics course were recorded in the file: Scores.mtp.

Follow these steps to construct a frequency table for the ages of students:

1. Open the worksheet.
   Choose **File>Open Worksheet**. Select the file A:Scores.mtp. Choose **Open**.

2. Construct the frequency table.
   Choose **Stat>Tables>Tally...** Place TestScore in the Variables: text box. Place checks in the Counts, Percents, Cumulative counts, and Cumulative percents checkboxes. Choose **OK**.

The Minitab Output

Tally for Discrete Variables: TestScore

```
TestScore  Count  CumCnt  Percent  CumPct
       60      2       2     3.08    3.08
       63      3       5     4.62    7.69
       66      1       6     1.54    9.23
       69      5      11     7.69   16.92
       72      9      20    13.85   30.77
       75     16      36    24.62   55.38
       78     14      50    21.54   76.92
       81      7      57    10.77   87.69
       84      6      63     9.23   96.92
       94      2      65     3.08  100.00
       N=     65
```

Figure 2.4

The section of the session window, as shown in Figure 2.4, labeled Tally for Discrete Variables: TestScore indicates the scores of the students, the count (frequency), the cumulative count (cumulative frequency), the percent (relative frequency) and the cumulative percent in each category.

Continuous data

The difficulty with continuous data, such as observations with temperatures or times, is that there may be no natural categories. There are several ways to deal with this problem. One way is to let Minitab determine the intervals for the categories and then read the histogram to find the corresponding frequencies for each interval.

Class Intervals

There are no strict guidelines for selecting either the number of class intervals or the interval lengths. Using a few relatively wide intervals will bunch the data, whereas using a great many relatively narrow intervals may spread the data over too many intervals, so that no interval contains more than a few observations. Generally speaking, $\sqrt{\text{number of observations}}$ often gives a rough estimate for an appropriate number of intervals. In the situation where the intervals (class widths) are equal, for either discrete or continuous data, frequencies or relative frequencies may be used on the vertical axis. However, if the intervals (class widths) are unequal, the vertical axis should be a density scale as indicated in the text in Section 2.4. You should choose the option of a Density histogram within Minitab.

Exercises

2.5 Adolescents with emotional problems are in a variety of environments. The family composition of 250 adolescents with emotional problems are stored in the file a:family.mtp. Construct a frequency table for the family composition of these adolescents containing counts, percents, cumulative counts, and cumulative percents. Which family composition is the most typical family composition for these adolescents? Which family composition is the least typical family composition for these adolescents?

2.6 The results of a poll of 375 young adults concerning their interest in volleyball are contained in the file a:vball.mtp. (Observe that the data is sorted.) Construct a frequency table for the categories containing counts, percents, cumulative counts, and cumulative percents. What is the percentage of young adults very interested in volleyball?

2.5   Dotplots and Histograms

New Minitab Commands

1. **Graph>Character Graphs>Dotplot** - Produces a character-based dotplot in the Session window. In this section, you will construct a dotplot for a large data set.

2. **Graph>Chart** - Produces many kinds of charts, including bar charts, line charts, symbol charts, and area charts.

   a. **Options** - By choosing appropriate items in the options category, you can create a variety of charts and transpose X and Y. In this section, you will place a check in the Transpose X and Y checkbox to interchange the variables defining the vertical and horizontal axes.

 b. **Edit Attributes** - Provides the opportunity to change the default appearance (colors, patterns, size, etc.) of the element in the active row in the Data display table. In this section, you will change the colors of the bars in the graph.

 c. **Annotation>Title** - Creates titles for a graph, placed at the top of the figure region, above the data region. You can create as many titles as you want, specifying different attributes for each title in the corresponding attributes rows (font, color, size, and so on). In this section, you will place a first title on the bar graph.

3. **Graph>Histogram** - Produces a histogram. Separates the data into intervals on the x-axis, and draws a bar for each interval whose height, by default, is the number of observations (or frequency) in the interval. In this section, you will construct a histogram for a large data set.

 a. **Options** - Contains options specific to Histogram. In this section, you will construct histograms with the percent (relative frequency) scale on the y-axis by darkening the option button for the Percent Type of histogram. You will also construct a cumulative percent (cumulative relative frequency) histogram with a cumulative percent scale on the y-axis by darkening the option button for the Cumulative Percent Type of histogram.

 b. **Frame>Ticks** - Controls the number of ticks, tick placement, tick appearance, and tick labels and their attributes. In this section, you will change the number of ticks on the Y axis.

 c. **Annotation>Title** - Creates titles for the histogram, placed at the top of the figure region, above the data region. You can create as many titles as you want, specifying different attributes for each title in the corresponding attributes rows (font, color, size, and so on). In this section, you will place a first title on the histogram.

Appropriate graphical representations of data often have more impact and convey more information quickly than a numerical summarization.

Dotplots
Dotplots are an attractive summary of numerical data when the data set is reasonably small or there are relatively few distinct data values.
The Problem - Data Temperatures on the Internet
The temperatures for a number of U.S. cities are available on the web. The temperatures for San Francisco, Ca. for the year 1999 are recorded in the file A:SanFrancisco.mtp. (http://www.engr.udayton.edu/weather/citylist.htm)

Follow these steps to construct a dotplot for the temperatures for 1999:

1. Open the worksheet.
Choose **File>Open Worksheet**. Select the file a:temps.mtp. Choose **Open**.

2. Construct the dotplot.
Choose **Graph>Character Graphs>Dotplot...** Place Temps in the **V**ariables: text box. Choose **O**K.

The Minitab Output

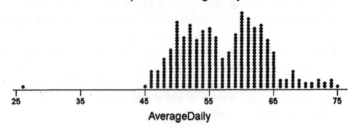

Figure 2.5

The section of the session window, as shown in Figure 2.5, labeled Dotplot for AverageDaily(Temperature) of the observations indicates the general distribution of the temperature and indicates one outlier.

Histograms (Bar Charts) for Categorical Data
Histograms that summarize categorical data are also referred to as bar graphs or as bar charts. These histograms for categorical data show the frequency corresponding to each category as a proportionally sized rectangular areas. Minitab calls to this type of histogram a chart. Minitab enables you to create simple bar charts, clustered bar charts, stacked bar charts, increasing bar charts, decreasing bar charts, cumulative bar charts, percent bar charts (by category), transposing axes or horizontal bar charts by choosing the appropriate items in the <Options> dialog box. Let's look at the following problem to construct a horizontal bar chart (histogram).

The Problem - Year in College
A sample of student responses at a small college to a questionaire provided the response as to the student's year in college.
Follow these steps to construct a histogram.

1. Open the worksheet.
   Choose **File>Open Worksheet**. Select the file a:students.mtp. Choose **Open**.

2. Construct the histogram.
   Choose **Graph>Chart**. Place Year in College in the first row of the X Graph variables: text box.
   a. Make the histogram a horizontal histogram, with the labels of the Year in College on the vertical axis.
      Choose **Options**. Place a check in the Transpose X and Y checkbox. Choose **OK**.
   b. Add Colors.
      Choose **Edit Attributes...**
      Choose **Fill Type>Solid**. Choose **Back Color>Blue**. Choose **OK**.
   c. Add a Title.
      Choose **Annotation>Title**. Place the title: Numbers of Students in Each

Year as the first title in Title text box. Choose **OK**. Choose **OK**.

The Minitab Output

Number of Students in Each Year

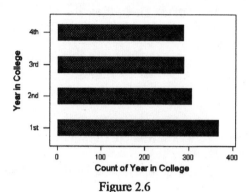

Figure 2.6

The histogram in Figure 2.6 shows each category represented by a bar.

The Problem - Toxic and Carcinogenic Chemicals
Sources of measurable atmospheric levels of toxic and carcinogenic chemicals con-
tribute to the total inhalable particulate matter (IPM). The frequency (number of
days) of measureable levels from a large metropolitan area are represented in the
following table.

Source	Frequency
Secondary Sulfate	228
Industry	25
Soil Resuspension	73
Motor Vehicles	34
Oil Burning	30
Incineration	38

Follow these steps to construct a histogram for this categorical data.
  a. Enter the data.
     Enter the labels: Sulfate, Industry, Soil, Motor Vehicles, Oil Burning, and
     Incineration in column C1. Name column C1 as Source. Enter the corre-
     sponding frequencies in column C2. Name column C2 as Frequency.
  b. Construct the histogram.
     Choose **Graph>Chart**. Place Frequency in the Y **G**raph variables: text
     box. Place Source in the X **G**raph variables: text box.
       1. Make the histogram a horizontal histogram, so that the long labels of
          the sources of the automobiles are legible.
          Choose **Options**. Place a check in the **T**ranspose X and Y checkbox.
          Choose **OK**.

2. Add Colors.
   Choose **Edit Attributes...** Choose **Fill Type>Solid**. Choose **Back Color>Blue**. Choose **OK**.

3. Add a Title.
   Choose **Annotation>Title**. Place the title: Sources of Toxic and Carinogenic Chemicals as the first title in Title text box. Choose **OK**. Choose **OK**.

The Minitab Output

Sources of Toxic and Carinogenic Chemicals

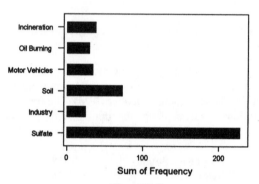

Figure 2.7

The histogram in Figure 2.7 shows each category represented by a bar.

Histograms for Numerical Data

A histogram for numerical data conveys the same sort of information contained in a stem-and-leaf display.

Frequency Histograms - Russian River Flow Data

The Russian River, located in California, is closely monitored at Healdsburg, Ca., for the rate of flow, measured in cubic feet per second (cfs). Data obtained from January 1, 1993 to March 18, 2000, was sorted into columns, one of which contains rate of flow less than 1000 cfs.

Follow these steps to construct a frequency histogram for the variable cfs<1000 of the rates of flow:

1. Open the worksheet.
   Choose **File>Open Worksheet**. Select the file a:russian.mtp. Choose **Open**.

2. Construct the frequency histogram.
   Choose **Graph>Histogram...** Place cfs<1000 (C3) in the first row of the X Graph variables: text box. Choose **OK**.

The Minitab Output

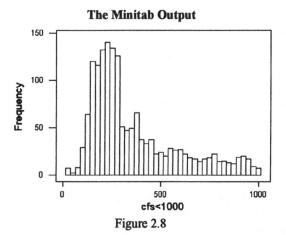

Figure 2.8

The graph window, as shown in Figure 2.8, indicates the general shape of the rate of flow (cfs) and the frequency of the rate of flow based upon the actual values of the variable cfs<1000. This histogram could now be read to produce the frequency table for the continuous data - cfs<1000.

Relative Frequency Histograms - Russian River Flow Data.

Follow these steps to construct a relative frequency histogram for the variable cfs<1000 of the rates of flow:
1. Construct the relative frequency histogram.
   Choose **Graph**>**Histogram...** Place cfs<1000 in the first row of the X **Graph** variables: text box.
   Click on the **Options** button. Darken the option button for the Percent Type of histogram. Choose **OK**. Choose **OK**.

The Minitab Output

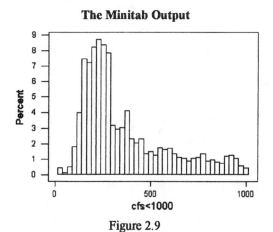

Figure 2.9

The graph window, as shown in Figure 2.9, indicates the relative frequency on

the vertical axis of the graph.
Cumulative Relative Frequency Histograms - Russian River Flow Data

Follow these steps to construct a cumulative relative frequency histogram for the variable cfs<1000 of the rates of flow:

1. Construct the cumulative relative frequency histogram.

   Choose **Graph>Histogram...** Place cfs<1000 in the Variables: text box. Click on the **Options** button. Darken the option button for the Cumulative Percent Type of histogram. Choose **OK**.

   a. Place additional tick marks on the axes.

   Choose **Frame>Ticks.** Place 11 in the X Number of Major text box (Counting from 0 to 1000, by 100's produces 11 tick marks.). Place 11 in the Y Number of Major text box (Counting from 0 to 100, by 10's produces 11 tick marks.). Choose **OK**.

   b. Add a title.

   Choose **Annotation>Title.** Place Cubic Rates of Flow < 1000 cfs in the first line of the Title text box. Choose **OK**. Choose **OK**.

The Minitab Output

Cubic Rates of Flow < 1000 cfs

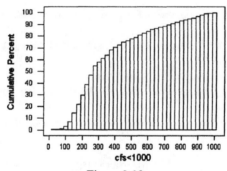

Figure 2.10

The graph window, as shown in Figure 2.10, indicates the cumulative relative frequency on the vertical axis of the graph and as a result of adding the tick marks, the graph can be more readily read and interpreted.

Exercises

2.7 The sodium content of a number of different "breakfast toaster bars" is provided in the following table:

190	210	220	200	205	195	215	225	205	210
185	225	195	210	190	190	225	200	220	195
190	240	225	212	210	195	230	230	218	215
200	205	205	220	220	210	240	210	225	225
240	215	210	240	240	230	210	215	235	240

Construct a dotplot of the sodium content of the "breakfast toaster bars". Describe the shape of the data set.

2.8 A survey of customer-service ratings indicated the following attributes and the frequency with which the attribute was rated as most important:

Attribute	Frequency
Sensitivity	173
Knowledge	236
Communication	405
Decisiveness	319
Efficiency	272

Construct a histogram for this categorical data. Include an appropriate title.
   a. Which source contributes the most to a high customer-service rating?
   b. What percentage of the total ratings are attributable to knowledge?

2.9 A retail store recently kept track of the type of camera purchased by customers with the last month:

Camera Type	Frequency
Camcorder	235
35mm	357
Instant	203
Other	40

Construct a histogram for this categorical data. Include an appropriate title.

2.10 Students indicated the times available for class in a survey at a local college:

Time	Frequency
Early morning	56
Mid-morning	93
Afternoon	46
Evening	27

Construct a histogram for this categorical data. Include an appropriate title.

2.11 The control of blood sugar levels is important to the health of diabetics. The levels of control of blood sugar levels were obtained from a clinic that treats diabetics and are stored in the file a:blood.mtp.
Follow these steps to construct a frequency histogram for the variable Hemoglobin:
   a. Open the worksheet.
      Choose **File**>**Open Worksheet**. Select the file a:blood.mtp. Choose **OK**.
   b. Construct the frequency histogram.
      Choose **Graph**>**Histogram...** Place Hemoglobin in the Variables: text box. Choose Options. Darken the Midpoint option button for types of intervals. Darken the Midpoint/cutpoint positions: option button in the Definition of Intervals category. Place 5:10/0.5 in the Midpoint/cutpoint positions: text box. Choose **OK**.
   c. Add a title.
      Choose **Annotation**>**Title**. Place Frequency Histogram for Blood Sugar Levels in the first line of the Title text box. Choose **OK**. Choose **OK**.
   d. Construct a relative frequency and cumulative relative frequency histogram for the variable Hemoglobin.

2.12 The following amounts represent the fees charged by a local delivery service for small items within the city.

4.24	3.76	3.28	6.40	5.98	3.37	3.08	3.11	4.00	3.22
4.54	4.10	4.97	4.23	4.87	5.47	3.06	3.16	5.32	4.95
5.76	4.11	7.18	5.19	4.87	3.82	4.04	4.04	3.96	7.20
4.39	4.21	4.04	6.01	8.23	4.88	5.17	6.00	3.00	4.07
5.32	5.55	4.26	5.78	4.93	4.11	4.24	4.32	3.17	4.28

Construct a relative frequency histogram of the data using appropriate midpoint positions and an appropriate title.

2.6   Do Sample Histograms Resemble the Population Histogram?

New Minitab Commands

1. **Calc**>**Random Data** > **Sample From Columns** - Randomly samples rows from one or more columns. You can sample with replacement (the same row can be selected more than once), or without replacement (the same row is not selected more than once).

Sample data is typically collected in order to make inferences about a population. Correct conclusions depend upon the sample being representative of the population. One issue concerns the extent to which histograms based on different samples from the same population resemble one another. If two different sample histograms can be expected to differ from one antoher in obvious ways, then at least one of the histograms will differ substantially from the population histogram. Sampling variability is a central concept in statistics.

The Problem - Bus Drivers

(Example 3.19) A sample of 708 bus drivers employed by public corporations was selectd, and the number of traffic accidents in which each was involved during a 4-year period was determined ("Application of Discrete Distribution Theory to the Study of Noncommunicable Events in Medical Epidemiology," *Random Counts in Biomedical and Social Sciences*, G. P. Patil, ed. University Park, Pa.: Penn. State Univ. Press, 1970).

Although the 708 observations actually constitued a sample from the population of all bus drivers, here we will regard the 708 observations as constituting the entire population.

Follow these steps to construct a relative frequency population histogram for the variable Accidents.

1. Open the worksheet.
   Choose **File**>**Open Worksheet**. Select the file a:bus.mtp. Choose **Open**.

2. Construct the relative frequency population histogram.
   Choose **Graph**>**Histogram...** Place Accidents in the first row of the X Graph variables: text box.

2.6   Do Sample Histograms Resemble the Population Histogram?

a. Click on the **Options** button. Darken the option button for the Percent Type of histogram.

Since the number of accidents occurred over the interval from 0 to 11, we will select 12 intervals. Darken the option button for the Number of Intervals. Place 12 in the Number of Intervals text box. Choose **OK**.

b. Place additional tick marks on the axes.

Choose **Frame>Ticks.** Place 12 in the X Number of Major text box (Counting from 0 to 11, by 1's produces 12 tick marks.). Place 6 in the Y Number of Major text box (Counting from 0 to 25, by 5's produces 6 tick marks.). Choose **OK**. Choose **OK**.

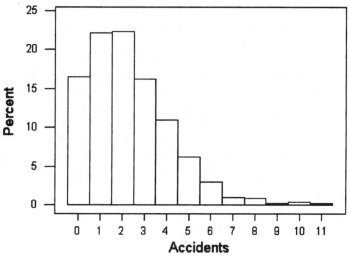

Figure 2.11

The graph window, as shown in Figure 2.11, indicates the population histogram for the number of accidents.

Follow these steps to construct relative frequency sample histograms of size 50 for the variable Accidents.

1. Select random samples of size 50 from the column Accidents.

Choose **Calc>Random Data > Sample From Columns.**

a. Select the first sample. Type 50 in the Sample text box. Place Accidents in the rows from column(s) text box. Place Sample1 in the Store samples in: text box. Choose **OK**.

b. Select the second sample. Type 50 in the Sample text box. Place Accidents in the rows from column(s) text box. Place Sample2 in the Store samples in: text box. Choose **OK**.

c. Select the third sample. Type 50 in the Sample text box. Place Accidents in the rows from column(s) text box. Place Sample3 in the Store samples in: text box. Choose **OK**.

d. Select the fourth sample. Type 50 in the Sample text box. Place Accidents in the rows from column(s) text box. Place Sample4 in the Store samples in: text box. Choose **OK.**

2. Construct the relative frequency sample histograms.

Choose **Graph>Histogram...**

a. Place Sample1 in the first row of the X Graph variables: text box.

b. Place Sample2 in the second row of the X Graph variables: text box.

c. Place Sample3 in the third row of the X Graph variables: text box.

d. Place Sample4 in the fourth row of the X Graph variables: text box.

e. Click on the **Options** button. Darken the option button for the Percent Type of histogram.

Since the number of accidents occurred over the interval from 0 to 11, we will select 12 intervals. Darken the option button for the Number of Intervals. Place 12 in the Number of Intervals text box. Choose **OK.**

f. Place additional tick marks on the axes.

Choose **Frame>Ticks.** Place 12 in the X Number of Major text box (Counting from 0 to 11, by 1's produces 12 tick marks.). Place 6 in the Y Number of Major text box (Counting from 0 to 25, by 5's produces 6 tick marks.). Choose **OK.** Choose **OK.**

One of the four graphs is shown in Figure 2.12. The histograms resemble one another in a general way, but there are some obvious dissimilarities. The population histogram rises to a peak and then declines smoothly, whereas the sample histograms tend to have more peaks, valleys, and gaps. Although the population data set contained an observation of 11, none of the four samples did.

The Minitab Output

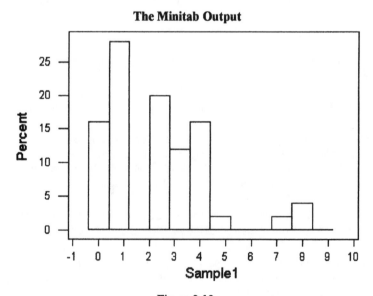

Figure 2.12

Chapter 3
Numerical Summaries

3.1   Overview

In the previous chapter you have examined graphical methods for displaying data. Although graphical methods provide a visual picture of the data, those graphical methods do not provide any numerical summary measures of the data. This chapter addresses the issue of providing numerical summaries. After reading this chapter you should be able to

1. Describe the Center and the Variability of a Data Set
2. Obtain numerical descriptive measures simultaneously for several variables
3. Construct a Boxplot

3.2   Measures of the Center and Variability of a Data Set

New Minitab Commands

1. **Stat>Basic Statistics>Display Descriptive Statistics** - Produces descriptive statistics (N, Mean, Median, Standard Deviation, etc.) for each variable or column. In this section, you will produce descriptive statistics for a small data set. **Graphs** option - Provides the option of displaying a histogram, a histogram with a normal curve, a dotplot, a boxplot, or a graphical summary of the variables. In this section, you will produce a Graphical summary. The Graphical summary checkbox option generates output in a Graph window. This graph includes a histogram with an overlaid normal curve, a boxplot, confidence intervals for the mean and median, and a table of statistics.

Numerical summaries that indicate where the center of a data set is located are called measures of central tendency. Measures of the center typically include the mean and median. Recall that the mean of a data set is the sum of the data divided by the number of pieces of data, while the median represents the middle value in an ordered data set and divides the data set into two equal parts.

Numerical summaries that describe the spread of values about the center are called measures of variability or measures of dispersion. Measures of variability typically include the range and standard deviation. The range represents the difference between the largest (maximum) and smallest (minimum) values in a data set. The standard deviation may be the most useful of all the measures of dispersion. The standard deviation is found by applying the equation

$$s = \sqrt{\frac{\sum x^2 - \frac{(\sum x)^2}{n}}{n-1}}$$

to the data.

Quartiles are a numerical summary that represent a measure of location. The lower quartile (Q1) represents the point such that 25% of the observations are below the point. The median is the second quartile (Q2) and is the point such that 50% of the observations are below the point. The upper quartile (Q3) represents the point such that 75% of the observations are below the point.

Let's look at the following problem and determine the numerical summaries for the data set.

The Problem - A Drought Watch

Recent lack of rainfall has a resulted in a drought declaration for the state of Pennsylvania for March, 2000. (http://marfchp1.met.psu.edu/Water/PHLESSCTP.html) The following table contains precipitation departures for the last 120 days preceding March 29, 2000:

```
Departure_from_Normal
    0.9   -2.5   -2.4   -3.6   -2.0   -1.6   -0.6   -1.6
    1.6    0.1    0.8   -2.0   -0.2   -0.8   -1.5   -2.1
    3.2    2.4    0.0   -2.5   -1.0    1.7    1.3   -0.6
   -1.3    1.4   -0.6   -1.0    1.8    0.0    0.2   -2.6
    2.7
```

Follow these steps to calculate the numerical summaries for the data set:

1. Enter Data.

   Enter the data into column C1. Name column C1 as Departure_from_Normal. As an alternative, open the file a:drought.mtp, by selecting **File>Open Worksheet**. Select the file a:drought.mtp. Choose **Open**.

2. Calculate the numerical summaries.

   Choose **Stat>Basic Statistics>Display Descriptive Statistics**. Place Departure_from_Normal in the Variables: text box. Choose **OK**.

The Minitab Output

Descriptive Statistics: Departure_from_Normal

Variable	N	Mean	Median	TrMean	StDev	SE Mean
Departur	33	-0.376	-0.600	-0.417	1.729	0.301

Variable	Minimum	Maximum	Q1	Q3
Departur	-3.600	3.200	-1.800	1.100

Figure 3.1

The Minitab output indicates the number of observations (N), the sample mean (Mean) the sample median (Median), the 5% trimmed mean (Tr Mean), sample standard deviation (StDev), and the standard error of the mean (SE Mean). The sample range is not listed, but can be obtained as Max-Min = 3.200 - -3.600 = 6.800. Measures of location include the first quartile (Q1) and the third quartile (Q3). (Note: Minitab uses a different algorithm than the text to determine quartiles. Quartiles determined using Minitab may not match exactly with your values obtained by using the method in the text.)

Minitab can also produce a graphical display of the data containing numerical summaries.

Follow these steps to produce a graphical display of the data containing the nu-

merical summaries.
  1. Calculate the numerical summaries.

Choose <u>S</u>tat><u>B</u>asic Statistics><u>D</u>isplay Descriptive Statistics. Place Departure_from_Normal in the <u>V</u>ariables: text box. Choose **Graphs..** Place a check in the <u>G</u>raphical summary checkbox. Choose **OK** . Choose **OK** .

The Minitab Output

Descriptive Statistics

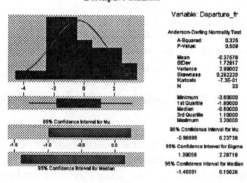

Figure 3.2

The Minitab output contains a histogram of the data set, a comparison with the normal curve, several confidence intervals, and various numerical summaries of the data set.

Exercises

3.1 Gasoline pumped at a local station has been tested for some time to determine if the average octane rating is 87.5. Random samples were collected and analyzed with the follow results:

86.2	87.4	88.3	87.1	88.3
88.5	87.1	86.3	86.8	97.4
86.5	87.4	87.2	88.7	87.2

Calculate the numerical summaries for the data set.
  a. Minitab calculates a trimmed mean (Tr Mean) by removing the smallest 5% and the largest 5% of the data values. In this data set does the Tr Mean differ much from the Mean?
  b. The observation 97.4 looks suspicious. Remove that outlier and determine the effect on the mean and median. Does the Mean change much with the removal of that observation? the median? the Tr Mean?

3.2 The ages of students in two sections of a statistics course were included in a survey conducted at the start of the class. The data is recorded in the file Ages.mtp. Calculate the numerical summaries for the data set.
  a. Comment on the relationship between the mean, median and trimmed mean.

3.3 A study investigating the effect of car speed on accident severity examined 50 accidents to determine the vehicle speeds at impact. The data is stored in the file a:speeds.mtp.

Chapter   3

a. Calculate the numerical summaries for the data set.

b. Obtain a Graphical summary of the data set.

c. Roughly what proportion of the speeds were between 28 mph and 46 mph?

d. Roughly what proportion of the speeds exceeded 64 mph (approximately 2 s.d. above the mean)?

3.4 Actual blood pressure values for 12 randomly selected individuals are as follows:

117.4	108.1	121.8	127.2	129.8	113.5
133.0	131.2	108.5	127.6	121.9	113.5

Calculate the numerical summaries for the data set.

a. Suppose that the third individual's blood pressure is 122.0 rather than 121.8. How does this change the median? What does this imply about the median's sensitivity to rounding or grouping of the data?

3.3   Obtaining Summary Measures for Several Variables

New Minitab Commands

1. **Stat>Basic Statistics>Display Descriptive Statistics** - Produces descriptive statistics (N, Mean, Median, Standard Deviation, etc.) for each variable or column. In this section, you will produce descriptive statistics for several variables in a large data set.

a. **By** variable: checkbox option - By placing a check in the **By** variable: checkbox you can display descriptive statistics separately for each value of a specified variable. In this section, you will produce descriptive statistics for both Females and Males in one summary table. If you also place a check in the Graphical summary checkbox multiple graphs are produced.

If a data set contains more than one variable, you can obtain the summary measures for a portion or all of the variables.

The Problem - Test Scores

High school students recently took exams designed to measure their knowledge in the areas of reading, writing, math, science and civics. These test scores are recorded in the file a:testscores.mtp.

Follow these steps to calculate the numerical summaries for the data set:

1. Open the worksheet.

Choose **File>Open Worksheet**. Select the file a:scores.mtp. Choose **Open**.

2. Calculate the numerical summaries.

Choose **Stat>Basic Statistics>Display Descriptive Statistics**. Place Reading, Writing, Math, Science and Civics in the Variables: text box. Choose **OK**

.

30

The Minitab Output

Descriptive Statistics

Variable	N	Mean	Median	Tr Mean	StDev	SE Mean
Reading	50	45.49	44.20	44.82	8.95	1.27
Writing	50	46.83	45.30	46.71	9.29	1.31
Math	50	45.15	44.00	44.59	7.81	1.10
Science	50	43.98	41.70	43.72	9.51	1.35
Civics	50	46.99	45.60	46.73	9.02	1.28

Variable	Min	Max	Q1	Q3
Reading	31.00	73.30	38.90	48.68
Writing	28.10	64.50	41.10	54.10
Math	32.70	68.00	39.28	49.05
Science	26.00	63.40	36.75	47.78
Civics	30.60	70.50	40.60	51.85

Figure 3.3

Again, the Minitab output indicates the number of observations (N), the sample mean (Mean) the sample median (Median), the 5% trimmed mean (Tr Mean), sample standard deviation (StDev), the standard error of the mean (SE Mean), the minimum value (Min). the maximum value (Max), the first quartile (Q1) and the third quartile (Q3) for all variables. You can identify that the Reading test scores have a sample mean of 45.49, a sample median of 44.20, a 5% trimmed mean of 44.82, a sample standard deviation of 8.92, a standard error of the mean of 1.27, a minimum of 31, a maximum of 73.3, a first quartile of 38.90, and a third quartile of 48.68.

You can choose the option of obtaining numerical summaries for the variables sorted by another variable. For example, you could obtain summary statistics for Reading sorted by Sex.

Follow these steps to calculate the numerical summaries for the variable Reading sorted by Sex:

1. Open the worksheet.

   Choose **File**>**Open Worksheet**. Select the file a:testscores.mtp. Choose **Open**.

2. Calculate the numerical summaries.

   Choose **Stat**>**Basic Statistics**>**Display Descriptive Statistics**. Place Reading in the Variables: text box. Place a check in the By variable: checkbox. Place Sex in the By variable: textbox. Choose **OK** .

The Minitab Output

Descriptive Statistics

Variable	Sex	N	Mean	Median	Tr Mean	StDev	SE Mean
Reading	Female	24	42.65	41.85	42.39	7.10	1.45
	Male	26	48.10	46.90	47.63	9.79	1.92

Variable	Sex	Min	Max	Q1	Q3
Reading	Female	31.00	60.10	36.95	46.90
	Male	34.20	73.30	41.60	52.10

Figure 3.4

The Minitab output indicates the number of Females (24), the mean of the Reading test scores (42.65) and other additional information. The number of Males is 26, while the mean of the Reading test score is 48.10. Other additional information is also provided.

Chapter 3

Exercises

3.5 Depression is associated with alterations in behavior and neuroendocrine systems that may be risk factors for decreased bone mineral density. Bone mineral density was measured at the femoral neck, spine, and at Ward's triangle for 12 women categorized as depressed and for 12 women categorized as normal. The file a:density.mtp contains the data set.

    a. Calculate the numerical summaries for the variables Spine, FemoralNeck, and WardsTriangle in the data set.

    b. Calculate the numerical summaries for the variables Spine, FemoralNeck, and WardsTriangle sorted by Status.

    c. Produce graphs containing the numerical summaries for each variable sorted by Status. Choose **Stat**>**Basic Statistics**>**Display Descriptive Statistics**. Place the variables Spine, FemoralNeck, and WardsTriangle in the Variables: text box. Place a check in the By variable: checkbox. Place Status in the By variable: textbox. Choose **Graphs..**. Place a check in the Graphical summary checkbox. Choose **OK**. Choose **OK**.

    d. Is depression a risk factor for decreased bone mineral density? Justify your answer by referring to the summary statistics and the graphs.

3.6 Socio-economic status and the type of school may be factors that influence mathematics test scores. The file a:math.mtp contains the data set.

    a. Calculate the numerical summary for the variable Math in the data set.

    b. Calculate the numerical summaries for the variable Math sorted by SocioEco.

    c. Calculate the numerical summaries for the variable Math sorted by SchType.

    d. Produce graphs containing the numerical summary for the variable Math sorted by the variable SocioEco.
    Choose **Stat**>**Basic Statistics**>**Display Descriptive Statistics**. Place the variable Math in the Variables: text box. Place a check in the By variable: checkbox. Place SocioEco in the By variable: textbox. Choose **Graphs..**. Place a check in the Graphical summary checkbox. Choose **OK**. Choose **OK**.

    e. Are there differences in math test scores by socio-economic status? Justify your answer by referring to the summary statistics and the graphs.

    f. Produce graphs containing the numerical summary for the variable Math sorted by the variable SchType.
    Choose **Stat**>**Basic Statistics**>**Display Descriptive Statistics**. Place the variable Math in the Variables: text box. Place a check in the By variable: checkbox. Place SchType in the By variable: textbox. Choose **Graphs..**. Place a check in the Graphical summary checkbox. Choose **OK**. Choose **OK**.

    g. Are there differences in math test scores by school type? Justify your answer by referring to the summary statistics and the graphs.

3.4   Constructing a Boxplot

New Minitab Commands

1. **Graph> Boxplot** - Produces a boxplot. A default boxplot consists of a box, whiskers, and outliers. Minitab draws a line across the box at the median. In this section, you will construct a boxplot from a data set.

    a. **Options** - Contains one options specific to Boxplot. You can transpose X and Y. Place a check in the Transpose checkbox to interchange the variables defining the vertical and horizontal axes.

The boxplot provides a quick display of some important features of the data. The boxplot "distills" the data set to its most important features and provides a formal tool for discriminating outliers during preliminary data analysis. Let's look again at the test scores as recorded in the file a:scores.mtp to construct a boxplot.

Reading Scores - a Boxplot

Follow these steps to construct the boxplot:

1. Open the worksheet.
   Choose **File**>**Open Worksheet**. Select the file a:testscores.mtp. Choose **Open**.

2. Construct the boxplot.
   Choose **Graph**> **Boxplot**. Place Reading in the Graph Variables: Y (measurement) text box. Choose **Options**. Place a check in the Transpose X and Y checkbox. Choose **OK**. Choose **OK**.

The Minitab Output

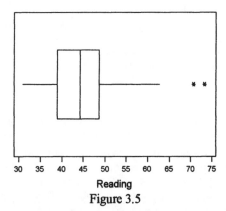

Reading
Figure 3.5

An examination of the boxplot indicates that a typical value is 44.20. The boxplot graphically depicts the position of the quartiles: Q_1, Q_2, Q_3 and indicates that 50% of the data fall between 38.90 (Q_1) and 48.68 (Q_3). The whiskers are the lines that extend from the left and right sides of the box to the adjacent values. The adjacent values in Minitab are the lowest and highest observations that are still inside the region defined by the following limits:

$$\text{Lower Limit} \quad Q_1 - 1.5(Q_3 - Q_1)$$
$$\text{Upper Limit} \quad Q_1 + 1.5(Q_3 - Q_1)$$

The left whisker indicates the minimum value (31.00). Observe that in this case there are two outliers recognized by Minitab (outliers will be identified by a *). Multiple boxplots for categories of a second variable can also be constructed. Let's again revisit the Reading test scores and construct boxplots for Reading for both Females and Males

Reading Scores - Side by Side Boxplots

Follow these steps to construct the boxplots:
1. Open the worksheet.
   Choose **File>Open Worksheet**. Select the file a:testscores.mtp. Choose **Open**.
2. Construct the boxplots.
   Choose **Graph> Boxplot**. Place Reading in the Graph Variables: Y (measurement) text box. Place Sex in the Graph Variables: X (measurement) text box. Choose Options. Place a check in the Transpose X and Y checkbox. Choose **OK**. Choose **OK**.

The Minitab Output

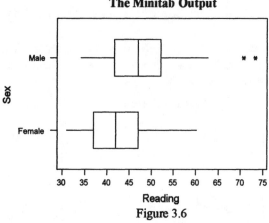

Reading

Figure 3.6

An examination of the boxplot for Females indicates that a typical value (Q_2) is 41.85. For Females the minimum (Min) value is 31, the first quartile (Q_1) is 36.95, the third quartile (Q_3) is 46.90 and the maximum (Max) value is 60.10. The boxplot graphically depicts the position of the quartiles: Q_1, Q_2, Q_3 and indicates that 50% of the data fall between 36.95 (Q_1) and 46.90 (Q_3).

An examination of the boxplot for Males indicates that a typical value (Q_2) is 46.90. For Males the minimum (Min) value is 34.2, the first quartile (Q_1) is 41.60, the third quartile (Q_3) is 52.10 and the maximum (Max) value is 73.3. The boxplot graphically depicts the position of the quartiles: Q_1, Q_2, Q_3 and indicates that 50% of the data fall between 41.60 (Q_1) and 52.10 (Q_3). Observe that in this case there are two outliers recognized by Minitab.

Exercises

3.7 Cholesterol counts of teenagers were examined where the teenagers were from two different socio-economic groups. The following sample data set is from that study.

Lower Socio-Economic Status

141	124	190	196	242	196	192	182	181	195	161
205	212	178	138	181	203	138	194	191	205	208
202	196	159	211	159	196	232	182	192	142	161
162	178	124	190	208	198	154	202	179		

Upper Socio-Economic Status

160	140	128	177	132	225	152	188	163	212	125
100	122	210	210	87	193	142	100	180	163	112
132	110	168	197	130	166	85	190	197	163	126
147	250	163	167	188	127					

a. Calculate the numerical summaries for the lower and upper socio-economic status groups, and comment.

b. Construct comparative stem-and-leaf displays for both groups.

c. Construct comparative boxplots.

3.8 Median prices of used homes sold in a number of metropolitan areas across the U.S. appeared in a recent edition of a national publication. (Prices have been rounded to the nearest $1000.)

85	76	89	114	111	98	126	60	89	121
95	89	179	91	69	80	85	134	98	84
85	112	114	91	94	81	105	261	136	101
78	103	108	86	98	99	77	86	89	77
88	112	112	94	79	148	91	117	74	372
135	177	77	77	94	117	113	97	115	74
77	104	94	112	83	78	103	81	97	140

a. Calculate the numerical summaries for the prices of used homes, and comment.

b. Construct a stem-and-leaf display for the prices of used homes.

c. Construct a boxplot for the prices of used homes.

3.9 The ages of individuals as they won a recent music award appear in the following table:

Males:	29	39	48	58	32	53	36	74	34	51
	39	29	37	57	35	46	45	37	42	38
	59	40	40	41	48	37	43	28	44	45

Females:	70	16	21	51	28	39	23	64	20	28
	23	31	33	31	32	35	16	31	24	43
	25	16	50	24	24	28	27	28	27	29

a. Calculate the numerical summaries for the males and females, and comment.

b. Construct comparative stem-and-leaf displays for both groups.

c. Construct comparative boxplots.

3.10 High temperatures (in degrees Fahrenheit) for August were recorded with the following results:

83	71	86	88	87	86	83	95	96	85
100	95	91	84	78	85	92	85	86	98
95	96	92	94	93	83	86	81	82	97
83									

a. Calculate the numerical summaries for the temperatures, and comment.

b. Construct a stem-and-leaf display for the temperatures.

c. Construct a boxplot for the temperatures.

Chapter 4
Summarizing Bivariate Data

4.1   Overview

This chapter introduces methods for describing relationships among various quantitative variables or characteristics, to predict a characteristic of interest, called the response or dependent variable. The characteristics used to predict the response variable are called the independent or predictor variables. After reading this chapter you should be able to

1. Construct a Scatter Plot
2. Obtain a Correlation Coefficeint Between Several Variables
3. Fit a Line to Bivariate Data
4. Assess the Fit of a Line
5. Fit a Straight Line to a (Transformed) Non-Linear Data Set

Relationships Among Data

You can develop models, which express the relationships among various characteristics, to predict a characteristic of interest, called the response or dependent variable. The characteristics used to predict the response variable are called the independent or predictor variables. Minitab will perform simple linear correlation(s), linear regression and multiple regression. Both numerical and graphical presentations are available.

4.2   Overview of Simple Linear Regression

New Minitab Commands

1. **Graph**>**Plot** - Produces a scatter plot that shows the relationship between two variables. In this section you will produce scatter plots for a number of variables.
   a. **Frame**>**Min** and Max - Defines, respectively, the minimum and maximum values for the scales in the data region. In this section you will construct a scatter plot where you have altered the minimum and maximum values on both the x and y axes.

2. **Stat**>**Basic Statistics**>**Correlation** - Calculates the Pearson product moment correlation coefficient between each pair of variables you place in the **Variables:** text box. In this section, you will use this command to determine Pearson's correlation coefficient.

3. **Manip**>**Rank** - Assigns rank scores to values in a column: 1 to the smallest value in the column, 2 to the next smallest, and so on. Ties are assigned the average rank for that value. In this section, you will use this command to determine the ranks in the calculation of Spearman's rank correlation coefficient.

Scatter Plots

A scatterplot is a graph of the relationship between the two characteristics of interest. The scatterplot provides a visual means of assessing the relationship between the variables and can assist us in proposing reasonable models.

The Problem

(Example 5.2, text) The growth and decline of forests is a matter of great concern in both the public and scientific communities. The paper "Relationships Among Crown Condition, Growth, and Stand Nutrition in Seven Northern Vermont Sugarbushes" (*Canad. J. of Forest Res.* (1995):386-397) indicates the percentage of mean crown dieback (y, the dependent variable), which is one indicator of growth retardation, and soil pH (higher pH indicates a more acidic soil) (x, the independent variable).

Soil pH	3.3	3.4	3.4	3.5	3.6	3.6	3.7	3.7	3.8	3.8
Dieback	7.3	10.8	13.1	10.4	5.8	9.3	12.4	14.9	11.2	8.0

Soil pH	3.9	4.0	4.1	4.2	4.3	4.4	4.5	5.0	5.1
Dieback	6.6	10.0	9.2	12.4	2.3	4.3	3.0	1.6	1.0

Follow these steps to construct a scatterplot of the data.

1. Enter data.

   Enter all of the observations to the right of the variable Soil pH in column C1. Name column C1 as Soil pH. Enter all of the observations to the right of the variable Dieback in column C2. Name column C2 as Dieback.

2. Save the data file.

   Choose **File>Save Current Worksheet As...** Select the drive a:. Position the cursor in the box labeled File Name: and type in the filename forest. Choose **Save**.

3. Create the scatterplot.

   Choose **Graph>Plot.** Place Dieback in the Y Graph variables: text box. Place Soil pH in the X Graph variables: text box.

4. Add a title.

   Choose **Annotation>Title.** Place the title: Mean Crown Dieback (%) vs. Soil pH in the first line of the Title text box.. Choose **OK.** Choose **OK.**

The Minitab Output

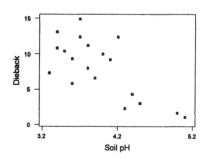

Figure 4.1

Large values of crown dieback tend to be associated with low soil pH - a negative or inverse relationship. The two variables appear to be approximately linearly related.

Exercises

4.1 The relationship between temperature (x, the independent variable) and the breaking strength of alloys (y, the dependent variable) has been examined. A study provided the following data.

Temperature	10	52	60	99	125	142
Breaking Strength	398	353	320	283	265	229

Temperature	152	180	225	265	280
Breaking Strength	155	120	63	38	18

  a. Construct a scatterplot of the variables.

  b. What do the plots suggest about the nature of the relationship between the variables?

4.2 The paper "Objective Measurement of the Stretchability of Mozzarella Cheese" (*J. of Texture Studies.* (1992) 185-194) reported on an experiment to investigate how the behavior of mozzarella cheese varied with temperature. Consider the accompanying data on x = temperature and y = elongation (%) at failure of the cheese. (Note: The researchers were Italian and used *real* mozzarella, not the poor cousin widely available in the United States.)

Temperature	59	63	68	72
Elongation	118	182	247	208

Temperature	74	78	83
Elongation	197	135	132

Follow these steps to construct a scatterplot of the data.

  a. Enter data.

    Enter all of the observations to the right of the variable Temperature in column C1. Name column C1 as Temperature. Enter all of the observations to the right of the variable Elongation in column C2. Name column C2 as Elongation.

  b. Create the scatterplot.

    Choose **Graph>Plot.** Place Elongation in the Y Graph variables: text box. Place Temperature in the X Graph variables: text box.

  c. Add a title.

    Choose Annotation>Title. Place the title: Elongation vs Temperature in the first line of the Title text box.. Choose **OK**. Choose **OK**.

  d. Construct a scatterplot with axes intersecting at (0,0).

    Choose **Graph>Plot.** Place Elongation in the Y Graph variables: text box. Place Temperature in the X Graph variables: text box.

    (i) Have the axes intersect at the point (0,0). Choose Frame>Min and Max. Darken the Separate minimum and maximum for X and Y axes option button. Place 0 in the Minimum for X: textbox. Place 100 in the Maximum for X: textbox. Place 0 in the Minimum for Y: textbox. Place 250 in the Maximum for Y: textbox. Choose **OK**.

  (ii) Mark the horizontal and vertical scales. Choose Frame>Tick. In order to mark 0, 20, 40, 60, 80, and 100 on the horizontal scale (6 points), Place 6 in the X Number of Major (tickmarks) textbox. In order to mark 0, 50, 100, 150, 200, and 250 on the vertical scale (6 points), Place 6 in the Y Number of Major (tickmarks) textbox. Choose **OK**. Choose **OK**.

  e. Construct a scatterplot with axes intersecting at (55,100).

     Choose **Graph>Plot.** Place Elongation in the Y Graph variables: text box. Place Temperature in the X Graph variables: text box.

    (i) Have the axes intersect at the point (55,100). Choose Frame>Min and Max. Darken the Separate mininimum and maximum for X and Y axes option button. Place 55 in the Minimum for X: textbox. Place 100 in the Maximum for X: textbox. Place 100 in the Minimum for Y: textbox. Place 250 in the Maximum for Y: textbox. Choose **OK**. Choose **OK**. (Observe that the position of the tick marks has changed.)

  f. Does the plot constructed with the axes intersecting at (0,0) or the plot constructed with the axes intersecting at (55,100) reveal more about the nature of the relationship?

  g. What do the plots suggest about the nature of the relationship between temperature and elongation? Justify your answer by referring to the scatterplots.

Correlation Coefficients

A scatterplot of bivariate numerical data gives a visual impression of the relationship between two variables. In order to make precise statements and draw conclusions from data, we need to go beyond pictures. A correlation coefficient is a quantitative assessment of the strength of a linear relationship between the ordered pairs of data.

Pearson's Sample Correlation Coefficient

Pearson's sample correlation coefficient is a popular measure of the strength of a linear relationship between two quantitative variables (where the bivariate data are assumed to be normally distributed).

The Problem

Age and right-hand grip strength were determined for a number of individuals. The data is in the following table:

Age (years)	16	17	19	12	17	23
GripStrength (lbs.)	58	58	65	53	60	68

Age (years)	18	26	13	15	26	24
GripStrength (lbs.)	65	72	55	58	69	72

Follow these steps to determine the correlation coefficient.

a. Enter data.

Enter all of the observations to the right of the variable Age in column C1. Name column C1 as Age. Enter all of the observations to the right of the variable GripStrength in column C2. Name column C2 as GripStrength.

b. Determine the correlation coefficient.

Choose **S̲tat**>**B̲asic Statistics**>**C̲orrelation**. Place Age and GripStrength in the **V̲**ariables: text box. Choose **OK**.

The Minitab Output

Correlations: Age, GripStrength

```
Pearson correlation of Age and GripStrength = 0.959
P-Value = 0.000
```

Figure 4.2

The correlation coefficient suggests that there is a strong positive linear relationship between age and grip strength. An examination of the scatterplot supports that idea of a strong positive linear relationship between the variables.

The Minitab Output

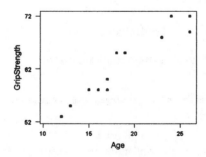

Age and Right-Hand Grip Strength

Figure 4.3

The scatter plot shown in Figure 4.3 suggests a moderate linear relationship.

Spearman's Rank Correlation Coefficient

Spearman's rank correlation coefficient (r_s) is not as sensitive as Pearson's correlation coefficient to outliers and identifies both linear and nonlinear relationships. Spearman's rank correlation coefficient is a measure of the correlation of the rankings of the data. Once you have determined the rankings of the pairs of data, Spearman's r_s is just Pearson's coefficient r applied to these rank pairs.

The Problem

Voice characteristics were scored on a scale from 0 to 6. The same conversation was scored by judges using (1) a taped conversation and (2) listening to the conversation on the phone.

Judge	Taped	Phone	Judge	Taped	Phone
A	3.26	1.47	F	4.63	2.52
B	5.04	2.52	G	3.15	1.37
C	4.62	2.21	H	4.09	1.68
D	3.99	1.37	I	4.12	2.02
E	2.33	1.72	J	2.50	1.98

Follow these steps to determine Spearman's rank correlation coefficient.

a. Enter data.

Enter all of the observations for the taped conversation in column C1. Name column C1 as Taped. Enter all of the observations for the phone conversation in column C2. Name column C2 as Phone.

b. Rank the data.

Choose **Manip>Rank.** Place Taped in the Rank data in: text box. Place RankTaped in the Store ranks in: text box. Choose **OK.** Choose **Manip>Rank.** Place Phone in the Rank data in: text box. Place RankPhone in the Store ranks in: text box. Choose **OK.**

c. Determine the correlation coefficient.

Choose **Stat>Basic Statistics>Correlation.** Place RankTaped and RankPhone in the Variables: text box. Choose **OK.**

The Minitab Output

Correlations: RankedTaped, RankedPhone

```
Pearson correlation of RankedTaped and RankedPhone = 0.683
P-Value = 0.030
```

Figure 4.4

This indicates a moderately positive relationship between the rankings of the scores of the judges.

Exercises

4.3 Severe acute adult malnutrition is common during a famine. A recent study focused on the use of the middle upper arm circumference (MUAC) as a measure to assess severe adult malnutrition rather than a more complicated traditional assessment, the body mass index (BMI) measurement.

MUAC	17.35	23.15	18.30	21.20	19.75	20.00
BMI	24.84	26.50	24.87	26.68	25.24	26.27

MUAC	18.21	23.87	19.64	22.69	20.92	19.85
BMI	24.78	30.76	24.87	28.65	28.90	26.19

a. Calculate the value of Pearson's correlation coefficient.

b. Construct a scatterplot of the variables.

c. Describe the relationship between the scatterplot and the correlation coefficient.

4.4 The shelf life of dry cereal is determined primarily by the moisture content. An examination of one particular cereal product indicated the following shelf lives (months) and moisture contents (%):

Shelf Life	0	4	8	10	12	14
Moisture Content	2.7	2.8	3.0	3.2	3.3	3.5

Shelf Life	16	18	20	22	24	26
Moisture Content	3.5	3.3	3.0	3.5	3.8	4.0

a. Calculate the value of Pearson's correlation coefficient.

b. Construct a scatterplot of the variables.

c. Describe the relationship between the scatterplot and the correlation coefficient.

4.5 The size of a three-bedroom home in 1000's of sq. ft. and the selling price ($1000) in an eastern city were recently examined.

Sq. ft.	2.3	1.9	2.7	3.0	2.5	2.2	2.9
Selling price	138	115	155	178	146	128	162

a. Calculate the value of Pearson's correlation coefficient.

b. Does it appear that the size of the home and the selling price are highly correlated?

4.6 Data relating the ages of wives (x = wife's age) and husbands (y = husband's age) at the time of marriage were obtained from city records:

Wife's age	18	23	41	32	30	25	17
Husband's age	22	27	53	31	37	29	22

a. Calculate the value of Pearson's correlation coefficient.

b. Construct a scatterplot of the variables.

c. Does it appear that the ages are positively or negatively correlated?

4.7 Is there a correlation between test anxiety and exam score performance? Data on x = score on a measure of test anxiety and y = exam score consistent with summary quantiles given in the paper "Effects of Humor on Test Anxiety and

Performance" (*Psych. Reports* (1999) 1203-1212) appears below.

TestAnxiety	23	14	14	0	7
ExamScore	43	59	48	77	50

TestAnxiety	20	20	15	21
ExamScore	52	46	51	51

a. Construct a scatterplot and comment on the features of the plot.

b. Does there appear to be a linear relationship between test anxiety score and exam performance? Based on the scatterplot, would you characterize the relationship as positive or negative? Strong or weak?

c. Calculate the value of Pearson's correlation coefficient. Is the value consistent with your answer to (b)?

d. Based on the value of the correlation, is it reasonable to conclude that test anxiety caused poor exam performance? Explain.

4.8 The age of the individual and the degree of exposure of that individual to hazardous chemicals were studied. The age of the individual and the mineral concentration in parts per million of the individual's tissue samples were recorded.

Age	50	72	62	81	35	76	23
Mineral Concentration	6	45	33	165	2	48	12

a. Calculate the value of Spearman's rank correlation coefficient.

b. Construct a scatterplot of the ranks.

c. Does it appear that the ranks are positively correlated?

4.3   Fitting a Line to Bivariate Data

New Minitab Commands

1. **Stat>Regression>Regression** - Performs simple, polynomial regression, and multiple regression using the least squares method.

   a. **Options** - Permits various options: weighted regression, fit the model with/without an intercept, calculate variance inflation factors and the Durbin-Watson statistic, and calculate and store prediction intervals for new observations. In this section, you will use this command to make predictions using the least squares regression line.

2. **Stat>Regression>Fitted Line Plot** - Fits a simple linear or polynomial (second or third order) regression model and plots a regression line through the actual data or the log10 of the data. The fitted line plot shows you how closely the actual data lie to the fitted regression line. In this section, you will obtain a fitted line plot to illustrate how the estimated relationship fits the data in a simple linear regression model.

Given two variables x and y, the general objective of regression analysis is to use information about x to make predictions concerning y. The roles played by the two variables are reflected in the terminology: y is referred to as the dependent or response variable, while x is referred to as the independent, predictor, or explanatory variable. We can model the response variable as a linear relationship of the independent variable. The simple linear regression model is a straight line of the form

$$y_i = a + bx$$

where   1. a is the y-intercept, the point on the y-axis where the straight line crosses the y-axis,

2. b is the slope, the amount by which y increases when x increases by 1 unit.

The Problem

(Example 5.9, text) Corrosion of the steel frame is the most important factor affecting the durability of reinforced concrete buildings. It is believed that carbonation of the concrete (caused by a chemical reaction that lowers the pH of the concrete) leads to corrosion of the steel frame and thus reduces the strength of the concrete. Representative data on x = carbonation depth (mm) and y = strength of concrete (Mpa) for a sample of core specimens taken from a particular building were read from a plot in the article "The Carbonation of Concrete Structures in the Tropical Environment of Singapore" (*Magazine of Concrete Research* (1996): 293-300).

Depth, x	8.0	20.0	20.0	30.0	35.0	40.0	50.0	55.0	65.0
Strength, y	22.8	17.1	21.5	16.1	13.4	12.4	11.4	9.7	6.8

Follow these steps to determine the least squares equation between carbonation depth and strength of concrete and to predict the strength of concrete when the carbonation depth is 25 (mm).

a. Enter data.

Enter the data for the carbonation depth in column C1. Name column C1 as Depth. Enter the data for the strength of concrete in column C2. Name column C2 as Strength.

b. Obtain the regression equation.

Choose **Stat>Regression>Regression.** Place Strength in the Response: text box. Place Depth in the Predictors: text box.

c. Make a prediction.

Choose **Options.** Place 25 in the Prediction intervals for new observations: text box. Choose **OK.** Choose **OK.**

The Minitab Output

Regression Analysis: Strength versus Depth

The regression equation is
Strength = 24.5 - 0.277 Depth

Predictor	Coef	SE Coef	T	P
Constant	24.517	1.079	22.73	0.000
Depth	-0.27694	0.02702	-10.25	0.000

S = 1.416   R-Sq = 93.8%   R-Sq(adj) = 92.9%

Analysis of Variance

Source	DF	SS	MS	F	P
Regression	1	210.67	210.67	105.03	0.000
Residual Error	7	14.04	2.01		
Total	8	224.72			

Predicted Values for New Observations

New Obs	Fit	SE Fit	95.0% CI	95.0% PI
1	17.593	0.556	(16.277, 18.910)	(13.993, 21.194)

Values of Predictors for New Observations

New Obs	Depth
1	25.0

Figure 4.5

The Minitab output, as shown in Figure 4.5, indicates the regression equation: Strength = 24.5 - 0.277 Depth. In the table just below the equation, appears information concerning the y intercept (Constant) and the slope of the variable Depth (carbonation depth). The column labeled "coefficient" indicates the values of the y-intercept (using more significant figures) is 24.517 and the slope is -0.27694. The additional information appearing below will be addressed at a later point, with the exception of the last line. The last line in the Minitab output indicates the predicted value (Fit) of y (Strenght) for a Depth (x) of 25 is 17.593.

A plot illustrating how the estimated relationship fits the data is possible in Minitab. This plot is called a fitted line plot.

Follow these steps to obtain a fitted line plot.

    a. Construct the fitted line plot.
        Choose **Stat**>**Regression**>**Fitted Line Plot**. Place Strength in the Response (Y): text box. Place Depth in the Predictor (X): text box. Choose Options. Place Carbonation Depth and Strength of Concrete in the Title: text box. Choose **OK**. Choose **OK**.

The Minitab Output

Carbonation Depth and Strength of Concrete

Strength = 24.5168 - 0.276940 Depth

S = 1.41631   R-Sq = 93.8 %   R-Sq(adj) = 92.9 %

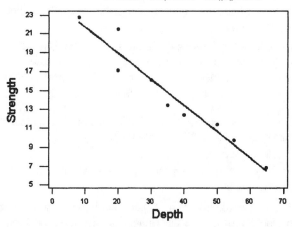

Figure 4.6

The fitted line plot, as shown in Figure 4.6, shows a scatterplot of the data with the least squares line superimposed upon the plot. The plot indicates a strong linear regression.

Exercises

4.9 It has been observed that Andean high-altitude natives have larger chest dimensions and lung volumes than do sea-level residents. Is this also true of lifelong Himalayan residents? The paper "Increased Vital and Total Lung Capacities in Tibetan Compared to Han Residents of Lhasa" (*Amer. J. Phys. Anthro.* (1991): 341-351) reported on the results of an investigation into this question. Included in the paper was a plot of vital capacity (y) versus chest circumference (x) for a sample of 16 Tibetan natives, from which the accompanying data was read.

Circumference (x)	79.4	81.8	81.8	82.3	83.7	84.3	84.3	85.2
Capacity (y)	4.3	4.6	4.8	4.7	5.0	4.9	4.4	5.0

Circumference (x)	87.0	87.3	87.7	88.1	88.1	88.6	89.0	89.5
Capacity (y)	6.1	4.7	5.7	5.7	5.2	5.5	5.0	5.3

a. Determine the least squares regression line and predict the vital capacity (y) when the chest circumference (x) is 80, 81, and 85 cm.

b. Obtain a fitted line plot.

c. Does the least squares line appear to give very accurate predictions? Explain your reasoning.

4.10 The size of a three-bedroom home in 1000's of sq. ft. and the selling price

($1000) in an eastern city were recently examined.

Sq. ft.	2.3	1.9	2.7	3.0	2.5	2.2	2.9
Selling price	138	115	155	178	146	128	162

a. Determine the least squares regression line and predict the Selling price when the Sq. ft. is 2.0, 2.4, and 2.8.

b. Obtain a fitted line plot.

c. Does the least squares line appear to give very accurate predictions? Explain your reasoning.

4.11 The relationship between the weight of a car (in pounds) and the fuel efficiency (in miles-per-gallon) has been under examination for some time.

Weight	3345	2540	2150	3225	3685	2400	3800
Mpg	25	34	43	28	23	36	22

a. Determine the least squares regression line and predict the fuel efficiency when the Weight is 2225, 2500 and 3000 pounds.

b. Obtain a fitted line plot.

c. Does the least squares line appear to give very accurate predictions? Explain your reasoning.

4.12 The following data resulted from an experiment in which weld diameter, x, and shear strength, y (lb), were determined for five different spot welds on steel. Mammalian hibernation has attracted a great deal of from physiologists and other scientists. A study focusing on the date hibernation began and the density of females in a particular species provided the following data.

Diameter (x)	200	210	220	230	240
Strength (y)	813.7	785.3	960.4	1118.0	1076.2

Follow these steps to determine the least squares equation between shear strength and weld diameter and to predict the strength when the weld diameter is 215.

a. Enter data.
Enter the data for the Diameter (x) in column C1. Name column C1 as Diameter. Enter the data for the Strength (y) in column C2. Name column C2 as Strength.

b. Obtain the regression equation.
Choose **Stat>Regression>Regression.** Place Strength in the Response: text box. Place Diameter in the Predictors: text box.

c. Make a prediction.
Choose Options. Place 215 in the Prediction intervals for new observations: text box. Choose **OK.** Choose **OK.**

d. Construct the fitted line plot.
Choose **Stat>Regression>Fitted Line Plot.** Place Strength in the Response (Y): text box. Place Diameter in the Predictor (X): text box. Choose Options. Place Shear Strength (y) versus Weld Diameter (x) in the Title: text box. Choose **OK.** Choose **OK.**

e. Since 1 lb = 0.4536 kg, strength observations can be reexpressed in kilograms through multiplication by this conversion factor: new y = 0.4536 (old y). **Follow these steps** to transform the data and determine the least squares equation between shear strength (y) and weld diameter (x), when y is expressed in kilograms.

(i) Transform the data.

Choose **Calc**>**Calculator**.   Place Kilograms in the <u>S</u>tore result in variable: text box. Click on the <u>E</u>xpression: text box. Type 0.4536 Expression: text box. Use the keypad to place the multiplication symbol in the <u>E</u>xpression: text box. Point to Strength in the <u>V</u>ariables list box and Select Strength, placing Strength in the <u>E</u>xpression: text box. The expression should now read as 0.4536 * 'Strength' in the <u>E</u>xpression: text box. Choose **OK**.

(ii) Obtain the regression equation.

Choose **<u>S</u>tat**>**<u>R</u>egression**>**<u>R</u>egression.** Place Kilograms in the <u>R</u>esponse: text box. Place Diameter in the Pre<u>d</u>ictors: text box.

(iii) Suppose that each y value in a data set is multiplied by a conversion factor c. What effect does this have on the slope, on the y intercept, and on the equation of the least squares line?

4.13 The following table indicate the disposable income and the amount of money spent on food per week for a family of four:

Disposable Income	21	32	24	36	26	28
Food Cost	83	110	52	115	88	80

a. Determine the least squares regression line and predict the Food Cost when the Disposable Income is 25, 30 and 35 (thousand) dollars.

b. Obtain a fitted line plot.

c. Does the least squares line appear to give very accurate predictions? Explain your reasoning.

4.4   Assessing the Fit of a Line

New Minitab Commands

1. **<u>S</u>tat**>**<u>R</u>egression**>**<u>R</u>egression** - - Performs simple, polynomial regression, and multiple regression using the least squares method.

a. **Graphs** - Displays residual plots. In this section, you will construct residual plots of the residuals versus the independent values.

Once the least squares line has been obtained, it is appropriate to ask how effectively the line summarizes the relationship between the dependent (y) and independent (x) variables. Specifically, we would like a quantitative indicator of the extent to which y variation can be attributed to the approximate linear relationship between the two variables.

The Coefficient of Determination

Variation in y can effectively be explained by an approximate straight-line relationship when the points in the scatter plot fall close to the least squares line. The difference between the observed values and their corresponding predictions obtained from the least squares line are termed residuals. A natural measure of the variation is the sum of the squared residuals:

$$SS_{Re\,sid} = \sum (y - \hat{y})^2 .$$

A second sum of squares assesses the total amount of variation in observed y values.

$$SS_{To} = \sum (y - \bar{y})^2 .$$

The coefficient of determination, denoted by , is given by .It is the proportion of variation in y that can be attributed to an approximate linear relationship between x and y in the sample.

Figure 4.5 displays a (partial) Minitab output from the regression of Expenditures on UnemploymentRate. The $SS_{Re\,sid}$ and SS_{To} appear in the SS column of the Analysis of Variance table. Thus,

$$r^2 = 1 - \frac{SS_{Re\,sid}}{SS_{To}} = 1 - \frac{7.8280}{16.8099} = .5343 .$$

Minitab reports the percentage of variation as R-Sq = 53.4%. That is, 53.4% of the variation in Expenditures can be attributed to an approximate linear relationship between Expenditures and UnemploymentRate. You should observe that r^2 is simply Pearson's sample correlation coefficient, r, squared. ($0.731^2 = 0.534$)

Plotting the Residuals

A residual plot is a plot of the $(x, residual)$ ordered pairs. A desirable plot is one that exhibits no particular pattern such as curvature. An examination of the residual plot after determining the least squares line effectively amounts to examining y after removing any linear dependence on x. Sometimes this examination may reveal the existence of a nonlinear relationship.

The Problem

(Problem 5.42, text) The decline of salmon fisheries along the Columbia River in Oregon has caused great concern among commercial and recreational fishermen. The paper "Feeding of Predaceous Fishes on Out-Migrating Juvenile Salmonids in John Day Reservoir, Columbia River" (*Trans. Amer. Fisheries Soc.* (1991): 405-420) gave the accompanying data on y = maximum size of salmonids consumed by a norther squaw fish (the most abundant salmonid predator) and x = squawfish length, both in mm.

Length (x)	218	246	270	287	318	344
Size (y)	82	85	94	127	141	157

Length (x)	375	386	414	450	468
Size (y)	165	216	219	238	249

Follow these steps to determine the least squares equation between Length (x) and

Size (y).

    a. Enter data.

       Enter the data for squawfish length (x) in column C1. Name column C1 as Length. Enter the data for the maximum size of salmonids in column C2. Name column C2 as Size.

    b. Obtain a fitted line plot.

       Choose **Stat**>**Regression**>**Fitted Line Plot.** Place Size in the Response (Y): text box. Place Length in the Predictor (X): text box. Select Options. Place Maximum Size of Salmonids vs. Squawfish Length in the Title: text box. Choose **OK.** Choose **OK.** The Fitted Line Plot is shown in Figure 4.7.

    c. Obtain the regression equation.

       Choose **Stat**>**Regression**>**Regression.** Place Size in the Response: text box. Place Length in the Predictors: text box.

    d. Predict the maximum size of salmonid consumed by a northern squaw fish of length 375 mm.

       Choose **Options.** Place 375 in the Prediction intervals for new observations: text box. Choose **OK.**

    e. Display the full table of fits and residuals.

       Choose **Results.** Darken the In addition, the full table of fits and residuals option button. Choose **OK.**

    f. Obtain a plot of the residuals.

       Choose **Graphs..** Darken the Regular option button for Residuals for Plots. Place Length in the Residuals versus the variables: textbox. Choose **OK.** Choose **OK.** The plot of the residuals is shown in Figure 4.8.

The Minitab Output

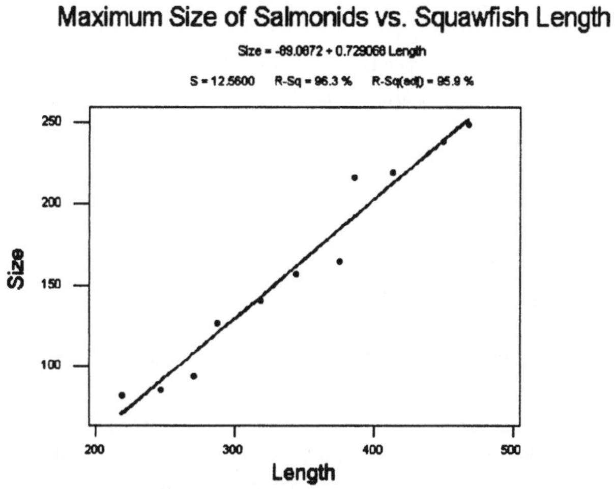

Maximum Size of Salmonids vs. Squawfish Length

Size = -89.0872 + 0.729068 Length

S = 12.5600   R-Sq = 96.3 %   R-Sq(adj) = 95.9 %

Figure 4.7

The fitted line plot, as shown in Figure 4.7, provides a visual verification of a reasonably substantial linear pattern, with the coefficient of determination at 96.3%. An examination of the Session window indicates the predicted maximum size of salmonid consumed by a northern squaw fish of length 375 mm to be 184.31 mm and the residual corresponding to the observation (375, 165) to be -19.31. The plot of the residuals, as shown in Figure 4.8, provides a visual assessment of the model effectiveness in making predictions.

The Minitab Output

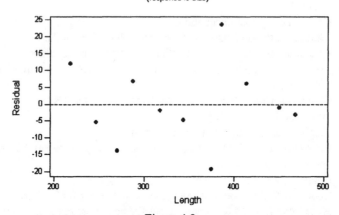

Figure 4.8

The residual plot shows no unusual pattern or discrepant values. The plot appears to support the appropriateness of a simple linear relationship.

The Problem

Growth patterns of children have been examined over a large period of time. The following data may provide some insight into the relationship between Age and Height.

Age (x)	2	3	4	5	6	7
Height (y)	34.1	37.7	40.6	43.3	45.9	48.2

Age (x)	8	9	10	11	12	13
Height (y)	50.5	52.8	55.2	55	57.2	59

Follow these steps to determine the least squares equation between Age (x) and Height (y).

    a. Enter data.

       Enter the data for Age in column C1. Name column C1 as Age. Enter the data for the Height in column C2. Name column C2 as Height.

    b. Obtain a fitted line plot.

       Choose **Stat>Regression>Fitted Line Plot.** Place Height in the

Response (Y): text box. Place Age in the Predictor (X): text box. Choose **OK.**

c. Obtain the regression equation.

Choose **Stat>Regression>Regression.** Place Height in the Response: text box. Place Age in the Predictors: text box.

d. Obtain a plot of the residuals.

Choose **Graphs..** Darken the Regular option button for Residuals for Plots. Place Age in the Residuals versus the variables: textbox. Choose **OK.** Choose **OK.**

The Minitab Output

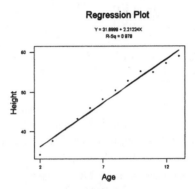

Regression Plot

Y = 31.8999 + 2.21224X
R-Sq = 0.978

Figure 4.9

The fitted line plot, as shown in Figure 4.9, suggests that the data is a fairly linear pattern with the coefficient of determination at 97.8%. The plot of the residuals, as shown in Figure 4.10, enables you to determine if the residuals have any pattern or structure and thus a visual assessment of the model effectiveness in making predictions.

The Minitab Output

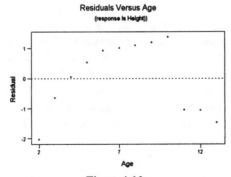

Residuals Versus Age
(response is Height)

Figure 4.10

The residual plot clearly shows structure in the residuals. The slight curvature in the original data has been magnified many times in this plot. The curvature is due

53

to the fact that the growth rate for children is not constant. In fact, younger children grow slightly faster than older children over this interval.

Exercises

4.14 (text 4.54) The paper "Aspects of Food Finding by Wintering Bald Eagles" (*The Auk* (1983):477-484) examined the relationship between the time that eagles spend aerially searching for food (indicated by the percentage of eagles soaring) and relative food availability.

Salmon (x)	0.0	0.0	0.2	0.5	0.5	1.0
Eagles (y)	28.2	69.0	27.0	38.5	48.4	31.1

Salmon (x)	1.2	1.9	2.6	3.3	4.7	6.5
Eagles (y)	26.9	8.2	4.6	7.4	7.0	6.8

a. Obtain a fitted line plot, making note of the coefficient of determination.

b. Obtain a residual plot of the standardized residuals vs fits. Does the residual plot suggest any structure in the residuals? That is, is there appear to be a nonrandom pattern to the residual plot?

4.15 Bicyclists are well aware that riding a bicycle poses considerable risks. The paper "Effects of Bike Lanes on Driver and Bicyclist Behavior" (ASCE *Trans. Eng. J.* (1977) 243-256) provided the following data on x = available travel space (distance between a cyclist and the roadway center line) and y = separation distance between a bike and a passing car.

TravelSpace	12.8	12.9	12.9	13.6	14.5
SeparationDistance	5.5	6.2	6.3	7.0	7.8

TravelSpace	14.6	15.1	17.5	19.5	20.8
SeparationDistance	8.3	7.1	10.0	10.8	11.0

a. Obtain a fitted line plot, making note of the coefficient of determination.

b. Obtain a residual plot of the standardized residuals vs fits. Does the residual plot suggest any structure in the residuals? That is, is there appear to be a nonrandom pattern to the residual plot?

4.16 (text, 5.52) The following data on x = frying time (sec) and y = moisture content (%) appeared in "Thermal and Physical Properties of Tortilla Chips as a Function of Frying Time" (*J. of Food Processing and Preservation* (1995): 175-189).

Time (x)	5	10	15	20	25	30	45	60
Moisture (y)	16.3	9.7	8.1	4.2	3.4	2.9	1.9	1.3

a. Obtain a fitted line plot, making note of the coefficient of determination.

b. Obtain a residual plot of the standardized residuals vs fits. Does the residual plot suggest any structure in the residuals? That is, is there appear to be a nonrandom pattern to the residual plot?

4.17 (text, 5.48) Anthropologists often study soil composition for clues as to how the land was used during different periods. The accompanying data on x = soil depth (cm) and y = percent montmorillonite in the soil was taken from a scatter plot in the paper "Ancient May Drained Field Agriculture: Its Possible Application Today in the New River Floodplain, Belize, C. A." (*Ag. Ecosystems and Environ.* (1984): 67-84)

Depth (x)	40	50	60	70	80	90	100
Percent (y)	58	34	32	30	28	27	22

a. Obtain a fitted line plot, making note of the coefficient of determination.

b. Obtain a residual plot of the standardized residuals vs fits. Does the residual plot suggest any structure in the residuals? That is, is there appear to be a nonrandom pattern to the residual plot?

4.5   NonLinear Relationships

New Minitab Commands

1. <u>Stat</u>><u>Regression</u>><u>Fitted Line Plot</u> - Fits a simple linear or polynomial (second or third order) regression model and plots a regression line through the actual data or the log10 of the data. The fitted line plot shows you how closely the actual data lie to the fitted regression line. In this section, you will obtain a fitted line plot to illustrate how the estimated relationship fits the data in a quadratic regression model.

When the ordered pairs in a scatter plot exhibit a linear pattern, it is relatively apparent that a straight line may provide a good fit to the points in the scatterplot. When a scatter plot shows a nonlinear pattern, finding an equation that fits the observed data may be more of a challenge.

An alternative to finding an equation that fits nonlinear data may be to transform x values and/or y values. One type of transformation that is useful for straightening a plot is a power transformation.

The Problem

The focus of many agricultural experiments is to study how the yield of a crop varies with the time at which it is harvested. The accompanying data appeared in the paper "Determination of Biological Maturity and Effect of Harvesting and Drying Conditions on Milling Quality of Paddy" (*J. of Agric. Engr. Research* (1975):353-361). The variables are x= time between flowering and harvesting (days) and y = yield of paddy, a type of grain farmed in India (kg/hectare).

Time (x)	16	18	20	22	24	26	28	30
Yield (y)	2508	2518	3304	3423	3057	3190	3500	3883

Time (x)	32	34	36	38	40	42	44	46
Yield (y)	3823	3646	3708	3333	3517	3214	3103	2776

Follow these steps to determine the least squares equation between Time (x) and Yield (y).

    a. Enter data.

       Enter the data for Time in column C1. Name column C1 as Time(x). Enter the data for the Yield in column C2. Name column C2 as Yield(y).

    b. Obtain a scatterplot.

       Choose **Graph>Plot.** Place Yield(y) in the Y Graph variables: text box. Place Time(x) in the X Graph variables: text box. Choose Annotation>Title. Place the title: Yield of Paddy vs Time in the first line of the Title text box.. Choose **OK**. Choose **OK.**

The Minitab Output

Yield of Paddy vs. Time

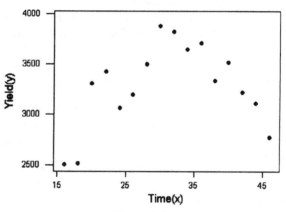

Figure 4.11

The scatterplot shown in Figure 4.11 appears to suggest a curved pattern, indicating that a linear relationship would not do a reasonable job of describing the relationship between Time (x) and Yield (y). In this case, the curved pattern seems to suggest a parabola (quadratic function).

    c. Fit a quadratic function to this data, obtaining a fitted line plot and storing the residuals.

       Choose **Stat>Regression>Fitted Line Plot.** Place Yield(y) in the Response (Y): text box. Place Time(x) in the Predictor (X): text box. Darken the Quadratic Type of Regression Model option button. Choose Storage. Place a check in the Residuals checkbox. Choose **OK**. Choose

<u>OK.</u>

The Minitab Output

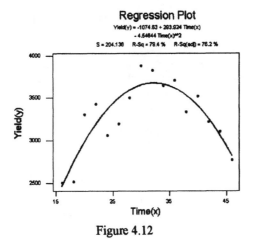

Figure 4.12

The Fitted Line Plot (Minitab output) is shown in Figure 4.12. The least squares quadratic that summarizes the relationship between Time(x) and Yield(y) is Yield(y) = -1074.63 + 293.924Time(x) - 4.54644Time(x)**2. This is equivalent to $Yield(y) = -1074.63 + 293.924x - 4.54644x^2$.

    d. Examine the Minitab output from fitting a quadratic function to the data. Choose <u>W</u>indow>Session.

The Minitab Output

Polynomial Regression Analysis: Yield(y) versus Time(x)

```
The regression equation is
Yield(y) = -1074.63 + 293.924 Time(x)
 - 4.54644 Time(x)**2


S = 204.138      R-Sq = 79.4 %      R-Sq(adj) = 76.2 %

Analysis of Variance
```

Source	DF	SS	MS	F	P
Regression	2	2086388	1043194	25.0334	0.000
Error	13	541738	41672		
Total	15	2628126			

Source	DF	Seq SS	F	P
Linear	1	197307	1.1364	0.304
Quadratic	1	1889081	45.3320	0.000

Figure 4.13

The Minitab output, as shown in Figure 4.13, indicates the regression equation and the coefficient of determination, $R^2 = 79.4\%$, indicating that 79.4% of the variability in yield may be explained by an approximate quadratic relationship between yield and time between flowering and harvesting.

   e. Obtain a residual plot for the quadratic regression between Time(x) and Yield(y).
   Choose **Graph>Plot**. Place RESI1 (the residuals) in the Y Graph variables: text box. Place Time(x) in the X Graph variables: text box. Choose Annotation>Title. Place the title: Residuals vs Time(x) in the first line of the Title text box.. Choose **OK**. Choose **OK**.

The Minitab Output

Residuals vs. Time(x)

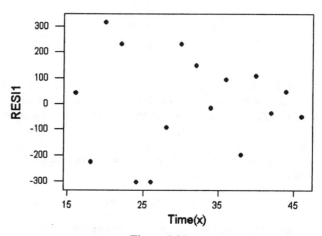

Figure 4.14

The plot of the residuals vs time is shown in Figure 4.14. The plot indicates no strong pattern in the residual plot for the quadratic.

Exercises

4.18 Tires that are underinflated or overinflated can increase tire wear and affect gas mileage. Tires were tested at different pressures to determine any influence on gas mileage.

Pressure (x)	29	31	32	33	34	35	36
Mpg (y)	30	33	35	36	35	32	27

   a. Construct a scatter plot of the data.
   b. Fit a quadratic function to this data, obtaining a fitted line plot and storing the residuals.
   c. Obtain a residual plot for the quadratic regression between Pressure(x) and Mpg(y).

4.19 The relationship between the number of accidents, per 50 million miles driven, and the driver's age in years is of interest to the insurance industry.

Accidents (y)	530	517	430	332	331	220	200	198
Age (x)	16	18	20	26	30	36	38	42

Accidents (y)	180	197	231	291	391	450	472
Age (x)	46	50	54	60	66	70	74

   a. Construct a scatter plot of the data..

   b. Fit a quadratic function to this data, obtaining a fitted line plot and storing the residuals.

   c. Obtain a residual plot for the quadratic regression between Age(x) and Accidents(y).

4.6 Transformations

New Minitab Commands

1. **Calc>Calculator** - Peforms arithmetic using an algebraic expression. You can use arithmetic operations, comparison operations, logical operations, functions, and column operations. In this section, you will use this calculator to transform data using natural logarithms, square roots, and exponents.

An alternative to finding a curve to fit the data is to find a way to "transform" the x values and/or y values so that a scatter plot of the transformed data has a linear appearance. Typical transformations may include taking square roots, logarithms or reciprocals of the data. The particular transformation that you select may on occasion be dictated by some theoretical argument.

The square root transformation tends to pull in the long tail of the distribution on the right (make large data values smaller), but stretch it out on the left (make small data values larger). On the other hand, a logarithmic transformation tends to make small data values larger and large data values smaller.

The Problem - Soggy Tortilla Chips

No tortilla chip lover likes soggy chips, so that it is important to find characteristics of the production process that produce chips with an appealing texture. The following data on x = frying time (sec.) and y = moisture content (%) appeared in the article "Thermal and Physical Properties of Tortilla Chips as a Function of Frying Time: (*J. of Food Processing and Preservation* (1995):175-189).

FryingTime(x)	5	10	15	20	25	30	45	60
Moisture(y)	16.3	9.7	8.1	4.2	3.4	2.9	1.9	1.3

Follow these steps to construct a scatter plot and obtain the regression line.

  a. Enter the data.

Enter the data for FryingTime in column C1. Name column C1 as FryingTime.

Enter the data for the Moisture in column C2. Name column C2 as Moisture.

b. Obtain a scatterplot.

Choose **Graph**>**Plot.** Place Moisture in the Y Graph variables: text box. Place FryingTime in the X Graph variables: text box. Choose Annotation>Title. Place the title: Moisture vs FryingTime in the first line of the Title text box.. Choose **OK.** Choose **OK.**

The Minitab Output

Moisture vs FryingTime

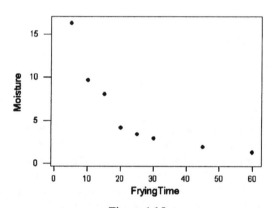

Figure 4.15

The scatterplot shown in Figure 4.15 appears to suggest an exponential decay, with the distribution of the original data being clearly positively skewed - a long upper tail to the right. One possibility might be to transform the y values (moisture content) by taking logarithms.

Follow these steps to transform the data and obtain a least squares regression line.

c. Transform the data.

Choose **Calc**>**Calculator.** Place LogMoisture in the Store result in variable: text box. Click on the down arrow for the Functions: drop down dialog box. Choose Logarithm. Choose Log 10. Double click on Moisture to replace number with Moisture in the LOGT function in the Expression: text box. Choose **OK.**

d. Obtain the regression equation.

Choose **Stat**>**Regression**>**Regression.** Place LogMoisture in the Response: text box. Place FryingTime in the Predictors: text box.

The Minitab Output

Regression Analysis: LogMoisture versus FryingTime

```
The regression equation is
LogMoisture = 1.14 - 0.0192 FryingTime

Predictor        Coef       SE Coef         T        P
Constant      1.14287       0.08016      14.26    0.000
FryingTi     -0.019170      0.002551     -7.52    0.000

S = 0.1246    R-Sq = 90.4%    R-Sq(adj) = 88.8%

Analysis of Variance

Source           DF           SS          MS        F        P
Regression        1       0.87736     0.87736    56.48    0.000
Residual Error    6       0.09320     0.01553
Total             7       0.97057
```

Figure 4.16

The Session window, as shown in Figure 4.16, indicates the resulting regression equation as $LogMoisture = 1.14287 - 0.019170 FryingTime$. To reverse this transformation, we take the antilog of both sides of the equation, obtaining: $y = 10^{(1.14287 - 0.019170 * FryingTime)}$

e. Obtain the fitted values.

Choose **C̲alc>Ca̲lculator.** Place FittedValues in the S̲tore result in variable: text box. Click on the down arrow for the F̲unctions: drop down dialog box. Choose Logarithm. Choose Antilog. Replace number with the right side of the logarithmic regression equation: 1.14287-0.019170 * 'FryingTime'. Choose **OK.**

f. Obtain a scatterplot of the original values and the fitted regression line.

Choose **G̲raph>P̲lot.** Place FittedValues in the first Y G̲raph variables: text box. Place FryingTime in the first X G̲raph variables: text box. Place Moisture in the second Y G̲raph variables: text box. Place FryingTime in the second X G̲raph variables: text box. Place Connect in the first Item Data Display textbox by selecting Connect from the Display drop down list box and selecting Graph from the For Each drop down list box. Place Symbol in the second Item Data Display textbox by selecting Symbol from the Display drop down list box and selecting Graph from the For Each drop down list box.
   a. Edit the attributes.

   Choose E̲dit Attributes. Place Solid in the first Graph text box by selecting Solid from the Line T̲ype drop down list box. Place None in the second Graph text box by selecting None from the Line T̲ype drop down list box. Choose **OK.**
   b. Place both graphs on the same page.

   Choose F̲rame>M̲ultiple Graphs. Darken the Ov̲erlay graphs on the same page option button. Choose **OK.** Choose **OK.**

The Minitab Output

Moisture vs FryingTime

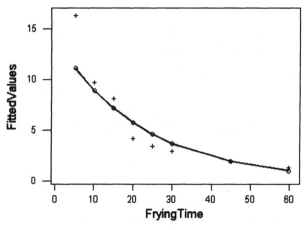

Figure 4.17

Exercises

4.20 Individuals were given training time (Training) before performing a certain repetitive operation (Performance). The data is as follows:

Training	20	20	35	35	50	50
Performance	5.3	6.0	3.8	4.8	3.9	3.2

    a. Construct a scatter plot of the data.

    b. Obtain the linear regression equation.

    c. Introduce transformations of Training:

        (i) $\sqrt{Training}$ and

        (ii) $\log_e Training$ (LOGE(Training)).

4.21 The relationship between addition of a substance (x) and its contribution to the strength (y) of the substance is of interest to the manufacturer.

Strength (y)	1.12	1.42	1.80	2.25	2.82	3.56	4.78	5.67	7.20
Content (x)	0.91	0.69	1.25	1.16	1.78	1.53	1.43	1.32	1.83

Strenght (y)	9.00	1.26	1.60	2.00	2.51	3.18	4.00	5.00	6.32
Content (x)	2.10	0.90	0.86	0.98	1.35	1.30	1.90	1.70	2.00

    a. Construct a scatter plot of the data.

    b. Introduce transformations.

(i) $\log_e$ *Strength* (LOGE(*Strength*)).

    c. Obtain the regression equation of $\log_e$ *Strength* on Content.

    d. Write the regression equation relating Permeability (y) to Content (x).

4.22 (5.54, Text) The paper "Asprects of Food Finding by Wintering Bald Eagles" (*The Auk* (1983): 477-484 examined the relationship between the time that eagles spend aerially searching for food (indicated by the percentage of eagles soaring) and relative food availability. The accompanying data is taken from a scatterplot that appeared in this paper. Let x denote salmon availability and y denote the percent of eagles in the air.

x	0.0	0.00	0.2	0.5	0.5	1.0
y	28.2	69.0	27.0	38.5	48.4	31.1

x	1.2	1.9	2.6	3.3	4.7	6.5
y	26.9	8.2	4.6	7.4	7.0	6.8

    a. Construct a scatter plot of the data.

    b. Introduce transformations.

      (i) $\sqrt{x}$ and $\sqrt{y}$.

      (ii) Construct a scatter plot of the transformed variables.

    c. Obtain the regression equation of $\sqrt{y}$ on $\sqrt{x}$.

    d. Obtain predicted values of y (Yhats) by using <u>C</u>alc>**Cal**cul**ator** and entering the square of the regression equation found in the previous step in the <u>E</u>xpression: text box.

    e. Obtain a scatterplot and compare this plot with the scatterplot of the original data.

Chapter 5
Probability

5.1　Overview

This chapter introduces the basic concepts of probability that are most widely used in statistics. Real data exhibit variability and that variability means uncertainty. Statistics uses probability to model the random behavior of real data as well as to quantify the uncertainty when inferences are made about a population of interest. Probability experiments via simulation may be performed a large number of times to understand underlying concepts of probability. Random samples with given probability distributions can be generated to examine and illustrate those probability distributions.

The relative frequency interpretation of probability holds that the probability of an event is the long-run proportion of the time that the outcome will occur. Minitab may be used to perform the calculations required in applying the relative frequency interpretation of probability. After reading this chapter you should be able to

　　　　1.　　Estimate Probabilities Using Simulation

5.2　Estimating Probabilities Using Simulation

New Minitab Commands

1. **Calc>Random Data>Integer** - Generates random data from an integer distribution, which is a discrete uniform distribution that ranges from the minimum to the maximum integer value specified. Each integer in the range has equal probability. In this section, you will use this command to simulate "guessing" on a true-false question.

2. **Stat>Tables>Tally** - Prints a one-way table of counts and percents for specified variables. In this section, you will use this command to obtain a table of the number of correct responses and the percentage of correct responses.

3. **Calc>Random Data>Sample From Columns** - Randomly samples rows from one or more columns. You can sample with replacement (the same row can be selected more than once), or without replacement (the same row is not selected more than once). In this section, you will use this command to sample data, without replacement, from one column.

Simulation provides a means of estimating probabilities that "generates observations" by performing an experiment that is similar in structure to the real situation of interest.

The Problem

A professor planning to give a quiz containing 20 true-false questions is interested in knowing how someone who answers by guessing will do in such a test. To investigate, he asks the 500 students in his introductory psychology course to write the numbers from 1 to 20 on a piece of paper and then to arbitrarily write T or F next to each number.

Follow these steps to simulate the experiment.

1. Generate random data.

   Choose **Calc>Random Data>Integer.** Place 500 in the Generate rows of data text box. Place C1-C20 in the Store in column(s): text box. Place 0 in the Minimum value: text box. Place 1 in the Maximum value: text box. Choose **OK.** (The 500 rows refers to the 500 students, while columns C1-C20 refers to the 20 true-false questions. The values of the random data are 0 and 1 where 0 denotes an incorrect response and 1 denotes a correct response.)

2. Obtain the score for each of the 500 students.

   Choose **Calc>Row Statistics.** Darken the option button for Sum. Place C1-C20 in the Input variables: text box. Place Scores in the Store result in: text box. Choose **OK.** (Here you are finding the sum of the 1's, the number of correct responses.)

3. Describe the scores.

   Choose **Stat>Tables>Tally.** Place Scores in the Variables: text box. Place a check in the Percents checkbox. Choose **OK.**

The Minitab Output

Summary Statistics for Discrete Variables

Scores	Percent
4	0.20
5	1.40
6	4.00
7	5.60
8	15.00
9	16.60
10	11.80
11	19.20
12	14.00
13	6.40
14	4.40
15	1.40

Figure 5.1

The Minitab output indicates the probability (percentage) of each Score occurring. The sample probability of obtaining a score of 10 or fewer correct items is 54.6%. This sample indicates a result reasonably consistent with pure chance (50%). Your results may be slightly different. Do you know why? This example illustrates how probability information can be used to make a decision and to get a sense of how likely it is that the decision is correct.

The Problem

Many California cities limit the number of building permits that are issued each year. One such city plan to issue permits for only ten dwelling units this upcoming year. The city decides who is to receive permits by holding a lottery. You are one of 39 individuals applying for permits. Thirty of these individuals are requesting permits for a single-family, eight are requesting a permit for a duplex, and one person is requesting a permit for a permit for a small apartment building with eight dwelling units. Each request is to be entered into the lottery. Requests will be selected at random one at a time, and if there are enough permits remaining, the request will be granted. This process will continue until all ten permits have been issued. If your request is for a single-family home, what are your chances of receiving a permit?

Follow these steps to simulate the experiment.

1. Create patterned data.
   Choose **Calc>Make Patterned Data > Simple Set of Numbers**. Place Requests in the Store patterned data in: text box. Place 1 in the From first value: text box. Place 39 in the To last value: text box. Choose **OK**.

2. Imitate the lottery drawing.
   Choose **Calc>Random Data>Sample From Columns**. Place 10 in from of rows in the Sample rows from columns: text box. Place Requests after columns: in the Sample rows from columns: text box. Place Permits in the Store samples in: text box. Choose **OK**.

3. Tally the results.
   Name column C3 as Number and column C4 as Total Number. By hand, enter the number of units issued for each request in the column Number and keep a running total on a seperate sheet of paper.

The Minitab Output

Requests	Permits	Number
1	26	1
2	32	2
3	28	1
4	18	1
5	24	1
6	35	2
7	9	1
8	12	1

Figure 5.2

The Minitab Data window indicates the results of one simulation of this lottery. If you were to repeat this process a large number of times, you would have an approximation to the probability of request number 1 being granted. We suimulated 500 such drawings and found that request number 1 was selected in 85 of the lotteries. Thus the estimated probability of receiving a permit $= \frac{85}{500} = .17$.

Exercises

5.1 Four students must work together on a group project. The responsibility for each part of the project is as follows:

Person:	Maria	Alex	Juan	Jacob
Task:	Survey design	Data collection	Analysis	Report Writing

.Because of the organization, one student must finish, before the next student can begin work. To ensure that the project is completed on time, a timeline is established, with a deadline for each team member. If any one of the team members is late, the timely completion of the project is jeopardized. Assume the following probabilities:

   a. The probability that Maria completes her part on time is 0.8

   b. If Maria completes her part on time, the probability that Alex completes on time is 0.9, but if Maria is late, the probability that Alex completes on time is 0.6.

   c. If Alex completes his part on time, the probability that Juan completes on time is 0.8, but if Alex is late, the probability that Juan completes on time is 0.5.

   d. If Juan completes his part on time, the probability that Jacob completes on time is 0.9, but if Juan is late, the probability that Jacob completes on time is 0.6.

   Use a four digit random number to represent each part of the project.
   Let the first digit represent Maria's part of the project with 1-8 representing on time and 0 and 9 representing late.
   Let the second digit represent Alex's part of the project where: if Maria was on time 1-9 represents Alex on time and 0 representing Alex being late; and if Maria was late 4-9 represents Alex on time and 0-3 represents Alex being late.
   Let the third digit represent Juan's part of the project where: if Alex was on time 2-9 represents Juan on time and 0-1 representing Juan being late; and if Alex was late 5-9 represents Juan on time and 0-4 represents Juan being late.
   Let the fourth digit represent Jacob's part of the project where: if Juan was on time 1-9 represents Jacob on time and 0 representing Jacob being late; and if Juan was late 3-9 represents Jacob on time and 0-2 represents Jacob being late.
   Thus, a number such as 3728 has:
   3 as the first digit representing Maria on time,
   7 as the second representing Alex on time, given that Maria was on time,
   2 as the third digit representing Juan on time, given that Alex on time, and
   8 as the fourth digit representing Jacob on time, given that Juan on time.
   This the entire project is completed on time.

   Follow these steps to simulate the experiment and determine the probability that the project is completed on time.

Chapter 5

a. Generate random data.

Choose **Calc>Random Data>Integer**. Place 20 in the Generate rows of data text box. Place Events in the Store in column(s): text box. Place 0000 in the Minimum value: text box. Place 9999 in the Maximum value: text box. Choose **OK**.

b. Print the Data window.

Choose **File>Print Window**. Choose **OK**.

c. Analyze the data.

Perform a analysis by hand consistent with the directions.

5.2 Using the results of the twenty trials from 5.1, determine the probability that the project is completed on time, given that: the probability that Maria completes her part on time is only 0.6. Let all other probabilities remain the same as in 5.1.

5.3 Many cities regulate the number of taxi licenses, and there is a great deal of competition for both new and existing licenses. Suppose that a city has decided to sell 10 new licenses for $25000 each. A lottery will be held to determine who gets the licenses, and no one may request more than 3 licenses. Twenty individuals and taxi companies have entered the lottery. Six of the 20 entries are requests for 3 licenses, 9 are requests for 2 licenses, and the rest are requests for a single license. The requests will be selected at random, filling as much of the request as possible.

a. Use simulation to determine the probability that a request for a single license will be granted.

b. Use simulation to determine the probability that a request for 3 licenses will be granted.

5.4 An individual claims to have the ability to predict the behavior of stocks on the stock market.

a. The probability of predicting the behavior of one particular stock on a Monday is 0.7.

b. If the behavior of the stock has been successfully predicted on Monday, the probability of predicting the behavior of the same particular stock on Tuesday is 0.8, but if the prediction is unsuccessful, the probability of predicting the behavior of the same particular stock on Tuesday is 0.6.

c. If the behavior of the stock has been successfully predicted on Tuesday, the probability of predicting the behavior of the same particular stock on Wednesday is 0.9, but if the prediction is unsuccessful, the probability of predicting the behavior of the same particular stock on Wednesday is 0.5.

d. If the behavior of the stock has been successfully predicted on Wednesday, the probability of predicting the behavior of the same particular stock on Thursday is 0.7, but if the prediction is unsuccessful, the probability of predicting the behavior of the same particular stock on Thursday is 0.2. Determine the probability that all predictions are successful.

Chapter 6
Population Distributions

6.1 Overview

This chapter begins to link together the basic concepts of probability with the concepts of statistical inference. This chapter introduces probability models that can be used to describe the distribution of characteristics of individuals in a population. Such models are essential if we are to reach conclusions based on a sample from the population. After reading this chapter you should be able to

1. Describe the Distribution of Values in a Population
2. Use a Continuous Probability Distribution as a Model for the Population Distribution of a Continuous Variable
3. Find Areas Under the Normal Curve
4. Construct a Normal Probability Plot

6.2 Describing the Distribution of Values in a Population

New Minitab Commands

1. **Calc>Make Patterned Data>Simple Set of Numbers** - Provides an easy way to fill a column with numbers that follow a pattern, such as the numbers 1 through 100, or five sets of 1, 2, and 3. This is very useful for entering a large number of patterned numbers. With this command, you can specify a pattern of equally spaced numbers, such as 10 20 30. In this section, you will enter a simple set of numbers.

2. **Calc>Calculator** - Does arithmetic using an algebraic expression.

   a. Partial sum (PARS) - Stores the sum of the first i rows of the input column in the ith row of the storage column. For example, if the first three rows of the input column contain the values 3, 5, and 2, then the first three rwos of the storage column will contain 3 (equal to 3+0), 8 (equal to 3+5) and 10 (equal to 8+2). In this section, you will use this command to calculate cumulative probabilities.

A population represents the entire collection of the variable(s) under consideration. These variables may be either categorical or numerical. For example, the variable *class standing* is categorical, with the four categories freshman, sophomore, junior and senior. The variable *time to complete registration*, measured in minutes, is numerical. The distribution of all the values of a numerical or categorical variable is called the population distribution.

Categorical Variables

The mode of transportation for each person entering the campus is recorded, with the following relative frequencies.

Mode	Relatvie Frequency
Driver	.529
Passenger	.090
Bike	.131
Bus	.110
Foot	.136
Skateboard	.004

Follow these steps to construct a bar chart.

1. Enter data.

    Enter the labels (Driver, Passenger, Bike, Bus, Foot, Skateboard) in column C1 in the Data window and the RelativeFrequecies in column C2. Name column C1 as Mode and column C2 as RelativeFrequency.

2. Create the bar chart.

    Choose **Graph>Chart.**

    Place RelativeFrequency in the Y Graph variables: text box. Place Mode in the X Graph variables: text box. Choose Annotation>Data Labels. Place a check in the Show data labels checkbox. Choose **OK**. The population distribution of modes of transportation is represented in the bar graph.

The Minitab Output

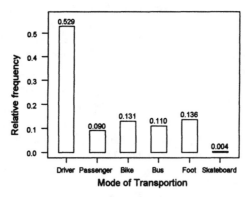

Figure 6.1

The population distribution of modes of transportation is represented in the bar graph shown in Figure 6.1. It is important to observe that the height of each rectangle corresponds to the relative frequency (proportion) for the corresponding value in the population. The graph indicates the probability of arriving by Bus is .110.

Numerical Variables

Numerical values may be considered to be either discrete or continuous. A discrete numerical variable is one whose possible values are isolated points on the number line. A continuous numerical variable is one whose possible values form an interval on the number line.

The population distribution for a discrete numerical variable can be summarized by a relative frequency histogram, whereas a density histogram is used to summarize the distribution of a continuous numerical value.

The Problem - a discrete numerical variable

The number of school-age children in each family in a sample of 100 families is contained in the file a:children.mtp.

Follow these steps to summarize the population distribution for the discrete numerical variable Children.

1. Open the worksheet.
   Choose **File>Open Worksheet**. Select the file a:children.mtp. Choose **Open**.

2. Construct the relative frequency histogram.
   Choose **Graph>Histogram...** Place Children in the Graph variables: text box. Choose Options. Darken the Percent option button for the Type of Histogram. Choose **Ok**. Choose **A**nnotation>**D**ata Labels. Place a check in the S**h**ow data labels checkbox. Choose **OK**. Choose **OK**.

The Minitab Output

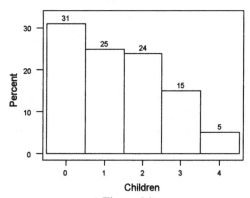

Figure 6.2

The population distribution of the number of school-age children is represented in the relative frequency histogram shown in Figure 6.2.

The Problem - a continuous numerical variable
    The amount of fill in a soft drink container in ounces is contained in the file a:fill.mtp.
Follow these steps to summarize the population distribution for the continuous numerical variable Fill.
  1. Open the worksheet.
     Choose **File>Open Worksheet**. Select the file a:fill.mtp. Choose **Open**.

  2. Construct the relative frequency histogram.
     Choose **Graph>Histogram...** Place Fill in the Graph variables: text box. Choose Options. Darken the Density option button for the Type of Histogram. (A density scale is placed on the y-axis. The total area under the histogram is one. The area of one bar is the proportion of the observations in that bar of the histogram.) Choose **OK**. Choose **OK**.

The Minitab Output

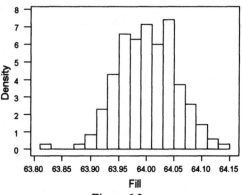

Figure 6.3

The population distribution of the amount of fill is represented in the density histogram shown in Figure 6.3.

Exercises
6.1 Of 150 hospital patients, 30% had type O blood, 38% had type A blood, 25% had type B blood, and 7% had type AB blood.
     a. Construct a bar chart of the categorical variable BloodType with the probabilities for each type indicated above each category.
     b. Use the information given in the bar chart, write a probability statement for the blood type AB.

6.2 The number of automobiles sold by each salesperson per week this last year is contained in the file a:sales.mtp.
     a. Construct a relative frequency histogram for the discrete numerical variable Sales with the probabilities indicated above each category.
     b. What is the probability of 3 or more cars being sold by one individual?

6.3 The rainfall in two particular counties in Pennsylvania for the month of June for the last 50 years is stored in the file a:rain.mtp.

  a. Construct a density histogram for the continuous random variable Rainfall

  .

  b. Summarize the data.
  Follow these steps to produce a graphical display of the data containing the numerical summaries.
  Choose **Stat**>**Basic Statistics**>**Display Descriptive Statistics**. Place Rainfall in the **V**ariables: text box. Place a check in the **B**y variable: checkbox. Place County in the **B**y variable: text box. Choose **Graphs..** Place a check in the **G**raphical summary checkbox. Choose **OK** . Choose **OK** .

  c. Compare the density histogram produced in (a) with the two graphical summaries produced in (b).

   (i) Is either distribution skewed or symmetric?

   (ii) Identify the mean and standard deviation of the Rainfall amounts for each county.

   (iii) Which county has the larger mean? the larger standard deviation?

6.4 Airlines frequently overbook flights. A flight with 100 seats has been booked with 110 reservations. Let x be the number of people who actually show up for a sold out flight. The population distribution of x is given in the following table.

x	95	96	97	98	99	100	101	102
Proportion	.05	.10	.12	.14	.24	.17	.06	.04

x	103	104	105	106	107	108	109	110
Proportion	.03	.02	.01	.005	.005	.005	.0037	.0013

   (i) What is the probability that the airline can accommodate everyone who shows up for the flight?

   (ii) What is the probability that not all passengers can be accommodated?

   (iii) If you are trying to get a seat on the flight and you are number 1 on the standby list, what is the probability, that you will get a seat on the flight?

Follow these steps to answer the above questions.

   (i) Enter the values of x.
   Choose **Calc**>**Make Patterned Data**>**Simple Set of Numbers.** Place Shows in the **S**tore patterned data in: text box. Place 95 in the **F**rom first value: text box. Place 110 in the **T**o last value: text box. Choose **OK**. (This command provides an easy way to fill a column with numbers that follow a pattern.)

   (ii) Enter Proportions.
   Enter the Proportions in column C1 in the Data window next to each corresponding value of x. Name column C2 as Proportions. (You have now set up the probability distribution of x.)

    (iii) Calculate cumulative proportions.

Choose **Calc**>**Calculator**. Place CumulativeProportions in the **S**tore result in variable: text box. Select Arithmetic from the **F**unctions: drop down dialog box. Choose Partial sum from the Arithmetic functions, placing PARS(number) in the **E**xpression: text box. Replace number in the PARS(number) in the expression with Proportions. Choose **OK**.

    (iv) Read the data window.

Read the values in the column CumulativeProportions to answer the above questions.

6.3   Population Models for Continuous Numerical Variables

Earlier you saw how a density histogram is used to summarize a population distribution when the variable of interest is numerical and continuous. The shape of the histogram and resulting probability statements that we may make based on the population distribution depend somewhat on the number and location of the intervals used in constructing the density histogram. The areas of the rectangles in such a histogram can be interpreted as the probability of observing a variable value in the corresponding interval. Specifically,

$$\text{density} = \frac{\text{relative frequency}}{\text{interval width}}.$$

The corresponding rectangle has height equal to density, so that the area of the rectangle is

$$\text{area} = (\text{height}) \cdot (\text{interval width}) = (\text{density}) \cdot (\text{interval width})$$
$$= \frac{\text{relative frequency}}{\text{interval width}} \cdot (\text{interval width}) = \text{relative frequency}$$

Thus the area of the rectangle above each interval is equal to the relative frequency of values that fall into the interval.

The Problem

The amount of fill in a soft drink bottle was measured, resulting in a sample of 120 observations of the variable

$$x = \text{amount of fill (oz.)}.$$

from the population of all amounts of fill of the soft drink bottles.

Follow these steps to summarize the population distribution for the continuous numerical variable AmountOfFill.

1. Open the worksheet.

Choose **File**>**Open Worksheet**. Select the file a:fill2.mtp. Choose **Open**.

2. Construct the relative frequency histogram.

Choose **Graph**>**Histogram...** Place AmountOfFill in the **G**raph variables: text box. Choose O**p**tions. Darken the **D**ensity option button for the Type of Histogram. Darken the option button for **C**utPoint Type of Intervals. Darken the option button for Mi**d**point/cutpoint positions:. Place Ounces in the Midpoint/cutpoint positions: text box. Choose **OK**. Choose **OK**.

The Minitab Output

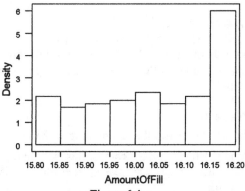

Figure 6.4

The resulting observations are summarized in the density histogram shown in Figure 6.4. The model can be used to approximate the probabilities involving the variable x. For example, the probability that the amount of fill is between 16 and 16.05 ounces is the area under the density curve and above the interval from 16 to 16.05 ounces. So P(amount of fill is between 16 and 16.05 ounces) = P(16<x<16.05) = 2·(.05)=.1 or 10%. Other probabilities are determined in a similar fashion.

Exercises

6.5 Consider the population of Lifetimes of light sensors made by a particular manufacturer.

Follow these steps to construct a population density curve of the probability distribution of the variable x = Lifetimes.

   a. Enter patterned data - the Lifetimes.

   Choose **Calc>Make Patterned Data>Simple Set of Numbers.** Place Lifetimes in the Store patterned data in: text box. Place 24.3 in the From first value: text box. Place 25.7 in the To last value: text box. Place .01 in the In steps of: text box. Choose **OK.** In this step, you are using this command to enter the values of the population of Lifetimes.

   b. Enter the density function.

   Choose **Calc>Calculator.** Place Density in the Store result in variable: text box. Select Arithmetic from the Functions: drop down dialog box. Choose Square root from the Arithmetic functions, placing SQRT(number) in the Expression: text box. Replace number in the SQRT(number) in the expression with 2/3.14159 - (Lifetimes - 25)**2. Choose **OK.** In this step, you are entering the density function for the Lifetimes of light sensors in preparation for construction of the density curve.

   c. Construct the density curve.

   Choose **Graph>Plot.**

   Place Density in the Y Graph variables: text box. Place Lifetimes in the X Graph variables: text box. Choose Connect from the Data display: Display drop down dialog box. Choose Frame>Tick. Place 24.3:25.7/.20 in the X Positions text box. Choose **OK**. Choose **OK**. In this step, you should now have the graph of a density function for a continuous numerical variable.

    d. Print the density curve.

Choose **File>Print Graph**. After selecting the correct printer, choose **OK** to print the density curve. Print 3 copies of the density curve.

    e. Shade the region under the curve corresponding to each of the following probabilities.

      (i) $P(24.5 < x < 24.9)$

      (ii) $P(x < 24.3)$

      (iii) The probability that the lifetimes exceed 25.1.

6.6 Consider the population of plastic components made by a particular manufacturer and define the variable x = breaking strength. The data is stored in the file a:strength.mtp.

    a. Construct a density histogram for the variable breaking strength. Use Cutpoints and let the Number of intervals be 10.

    b. Determine $P(246 < x < 258)$.

6.7 Consider the population that consists of all travel times of one individual to school. A reasonable model for the population is given by the following table:

Times	Density
0	0.0
10	0.5
20	0.0

Follow these steps to construct a density histogram of the probability distribution of the variable x = Times.

    a. Enter data.

Enter the Times in column C1 in the Data window and the Density in column C2. Name column C1 as Times and column C2 as Density.

    b. Create the density curve.

Choose **Graph>Chart**.

Place Density in the Y **G**raph variables: text box. Place Times in the X **G**raph variables: text box. Choose **C**onnect from the **D**ata display: Display drop down dialog box. Choose **OK**.

    c. Verify that the total area under the density curve is equal to 1.

    d. What is the probability that x is less than 10? less than 5? more than 15?

    e. What is the probability that x is between 5 and 10?

6.4   Normal Distributions

New Minitab Commands

1. **C**alc>**P**robability **D**istributions> **N**ormal - Allows you to calculate the probability densities, cumulative probabilities, and inverse cumulative probabilities for a normal distribution.For the continuous distributions, such as the normal distribution, Minitab calculates the continuous probability density function.

a. **Probability density:** In this section, you will darken the Probability density: option button in order to construct density curves of the normal probability distribution.

b. **Cumulative probability:** In this section, you will darken the Cumulative probability: option button in order to determine the area under the normal probability density function.

2. **Graph>Plot** - Produces a scatter plot that shows the relationship between two variables.

a. **Frame>Multiple Graphs** - In this section, you will darken the Overlay graphs on the same page option button to overlay graphs on the same page (in the same window).

3. **Calc>Standardize** - Centers and scales columns of data. In this section, you will use this command to determine standard scores, or z scores, for a column of data.

Normal distributions are continuous probability distributions. Normal distributions are frequently used as population models since normal distributions provide reasonable approximations to the distributions of many different variables. Different normal distributions are distinguished from one another by their mean, μ, and standard deviation, σ. The mean, μ, describes the center of the distribution, and the standard deviation, σ, describes the shape of the distribution.

Follow these steps to construct 3 density curves of the normal probability distribution.a. Enter patterned data.

Choose **Calc>Make Patterned Data>Simple Set of Numbers.** Place x in the Store patterned data in: text box. Place -10 in the From first value: text box. Place 100 in the To last value: text box. Place .1 in the In steps of: text box. Choose **OK**. In this step, you are setting the interval over which the values of x may occur.

b. Calculate densities.

(i) Density1.

Choose **Calc>Probability Distributions> Normal.** Darken the Probability density option button. Place 10 in the Mean: text box. Place 5 in the Standard deviation: text box. Darken the Input Column option button. Place x in the Input Column: text box. Place Density1 in the Optional storage: text box. Choose **OK**. In this step, you have entered the values of the density function for the normal distribution with a mean of 10 and a standard deviation of 5. These values necessary for the construction of the first density curve.

(ii) Density2.

Choose **Calc>Probability Distributions> Normal.** Darken the Probability density option button. Place 40 in the Mean: text box. Place 2.5 in the Standard deviation: text box. Darken the Input Column option button. Place x in the Input Column: text box. Place Density2 in the Optional storage: text box. Choose **OK**. In this step, you have entered the values of the density function for the normal distribution with a mean of 40 and a standard deviation of 2.5. These values necessary for the construction of the second density curve.

(iii) Density3.

Choose **Calc>Probability Distributions> Normal.** Darken the Probability density option button. Place 70 in the Mean: text box. Place 10 in the Standard deviation: text box. Darken the Input Column option button. Place x in the Input Column: text box. Place Density3 in the Optional storage: text box. Choose **OK**. In this step, you have entered the values of the density function for the normal distribution with a mean of 70 and a standard deviation of 10. These values necessary for the construction of the third density curve.

c. Graph the density curves.

Choose **Graph>Plot.**

Place Density1 in row 1 of the Y Graph variables: text box. Place x in the X Graph variables: text box. Place Density2 in row 1 of the Y Graph variables: text box. Place x in the X Graph variables: text box. Place Density3 in row 1 of the Y Graph variables: text box. Place x in the X Graph variables: text box. Choose Connect from the Data display: Display drop down dialog box. Choose Frame>Multiple Graphs. Darken the Overlay graphs on the same page option button under Generation of Multiple Graphs. Choose **OK**. Choose **OK**.

The Minitab Output

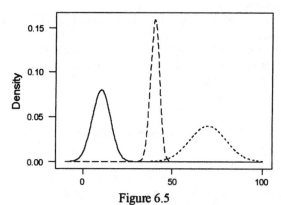

Figure 6.5

The resulting density curves, as shown in Figure 6.5, illustrate that the smaller the standard deviation, the taller and narrower the corresponding density curve.

The Standard Normal Distribution

The standard normal distribution is a normal distribution with

$$\mu = 0 \text{ and } \sigma = 1.$$

It is customary to let z represent a variable whose distribution is distribution is described by the standard normal curve. In working with the normal distribution, we need to be able to find areas under the standard normal distribution.

You can use Minitab to create a table of values from a normal probability distribution with a given mean and standard deviation. Typically, you have to convert a value of x to a z-value and then look up the answer in a table of areas under the normal curve in order to find a probability. Minitab provides a table of probabilities in terms of the original value, x. Minitab will enable you to determine cumulative areas to the <u>left</u> of a particular value of x, or a particular value of z.

The Problem- Looking up Areas under the Normal Curve

The area to the left of a z-value.

Follow these steps to determine the area under the normal curve to the left of z = -1.76.

1. Determine the area.

    Choose <u>C</u>alc><u>P</u>robability <u>D</u>istributions> <u>N</u>ormal. Darken the <u>C</u>umulative probability: option button. Place 0 in the <u>M</u>ean: text box. Place 1 in the <u>S</u>tandard deviation: text box. Darken the Input co<u>n</u>stant: option button. Place -1.76 in the Input co<u>n</u>stant: text box. Choose **O<u>K</u>**.

The Minitab Output

Cumulative Distribution Function

```
Normal with mean = 0 and standard deviation = 1.00000

        x       P( X <= x)
  -1.7600        0.0392
```

Figure 6.6

The resulting Minitab output, as shown in Figure 6.6, indicates the area to the left of z = -1.76 is .0392 or 3.92%.

The area to the right of a z-value.

Follow these steps to determine the area under the normal curve to the right of z = 0.58.

1. Determine the area.

    Choose <u>C</u>alc><u>P</u>robability <u>D</u>istributions> <u>N</u>ormal. Darken the <u>C</u>umulative probability: option button. Place 0 in the <u>M</u>ean: text box. Place 1 in the <u>S</u>tandard deviation: text box. Darken the Input co<u>n</u>stant: option button. Place 0.58 in the Input co<u>n</u>stant: text box. Choose **O<u>K</u>**.

The Minitab Output

Cumulative Distribution Function

```
Normal with mean = 0 and standard deviation = 1.00000

        x       P( X <= x)
   0.5800        0.7190
```

Figure 6.7

The resulting Minitab output, as shown in Figure 6.7, indicates the area to the left of z = 0.58 is 0.7190 or 71.9%. Since the area under the normal curve is 1 or 100%, the area to the right of z = 0.58 is $1 - 0.7910 = .2090$.

The area to the between two z-values.

Follow these steps to determine the area under the normal curve between z = 0.42 and z = 1.96.

Chapter   6

1. Determine the area to the left of z = 0.42.

   Choose Calc>Probability Distributions> Normal. Darken the Cumulative probability: option button. Place 0 in the Mean: text box. Place 1 in the Standard deviation: text box. Darken the Input constant: option button. Place 0.42 in the Input constant: text box. Choose OK.

2. Determine the area to the left of z = 1.96.

   Choose Calc>Probability Distributions> Normal. Darken the Cumulative probability: option button. Place 0 in the Mean: text box. Place 1 in the Standard deviation: text box. Darken the Input constant: option button. Place 1.96 in the Input constant: text box. Choose OK.

The Minitab Output

Cumulative Distribution Function

```
Normal with mean = 0 and standard deviation = 1.00000

         x       P( X <= x)
    0.4200         0.6628
```

Cumulative Distribution Function

```
Normal with mean = 0 and standard deviation = 1.00000

         x       P( X <= x)
    1.9600         0.9750
```

Figure 6.8

The resulting Minitab output shown in Figure 6.8 indicates the area to the left of z = 0.42 is 0.6628 or 66.28%; the area to the left of z = 1.96 is 0.9750 or 97.50%. The area between z = 0.42 and z = 1.96 is found by subtracting the two areas: .9750 − .6628 = .3122 or 31.22%.

Other Normal Distributions

You can calculate probabilities and describe values for any normal distribution. To calculate those probabilities, you may (a) use Minitab or (b)standardize the relevant values and use a table of areas under the normal curve.

If x is a variable whose behavior is described by a normal distribution with mean μ and standard deviation σ, then

$$P(x<b) = P(z<b^*)$$
$$P(a<x) = P(a^*<z) \quad P(x>a) = P(z>a^*)$$
$$P(a<x<b) = P(a^*<x<b^*)$$

where z is a variable whose distribution is standard normal and

$$a^* = \frac{a-\mu}{\sigma} \text{ and } b^* = \frac{b-\mu}{\sigma}$$

The Problem

In poor countries, the growth of children can be an important indicator of general levels of nutrition and health. Data in the paper "The Osteological paradox:Problems of Inferring Prehistoric Health from Skeletal Samples" (*Current Anthropology* (1992):343-370) suggests that a reasonable model for the population distribution of the continuous numerical variable x = height of a five-year old child is a normal distribution with mean $\mu = 100$ cm and standard deviation

$\sigma = 6$ cm.

Follow these steps to determine the proportion of the population that has a height between 94 cm and 112 cm.

1. Enter data.

   Enter the heights of 94 and 112 cm in column C1 in the Data window. Name column C1 as Heights.

2. Determine areas.

   Choose <u>C</u>alc><u>P</u>robability <u>D</u>istributions> <u>N</u>ormal. Darken the <u>C</u>umulative probability: option button. Place 100 in the <u>M</u>ean: text box. Place 6 in the <u>S</u>tandard deviation: text box. Darken the Input co<u>l</u>umn: option button. Place Heights in the Input column: text box. Place Areas in the Optional sto<u>r</u>age: text box. Choose <u>OK</u>.

The Minitab Output

↓	Heights	Area
1	94	0.158655
2	112	0.977250

Figure 6.9

The resulting Minitab output, as shown in Figure 6.9, indicates the area to the left of x = 94 (and z = -1.00) is 0.1587 and the area to the left of x = 112 (and z = -2.00) is 0.9773. The area between z = -1.00 and z = 2.00 is found by subtracting the two areas: .9773 − .1587 = .8186 or 81.86%.

Describing Extreme Values in a Normal Distributions

Self registration via computer is a popular method for completing registration for classes at many colleges. Data suggests that the distribution of the variable

$$x = \text{time logged on the computer}$$

can be approximated by a normal distribution with mean $\mu = 12$ minutes and standard deviation $\sigma = 2$ minutes. Some students do not log off the computer properly leading the college to decide to disconnect students automatically after some amount of time has elapsed. The time chosen is such that only 1% of the students will be disconnected while they are still attempting to register.

Follow these steps to determine the value of x^*, time logged on the computer before disconnecting the student.

1. Determine the value of x^*.

   Choose <u>C</u>alc><u>P</u>robability <u>D</u>istributions> <u>N</u>ormal. Darken the <u>I</u>nverse cumulative probability: option button. Place 12 in the <u>M</u>ean: text box. Place 2 in the <u>S</u>tandard deviation: text box. Darken the Input co<u>n</u>stant: option button. Place 0.99 (the area to the left of x^*) in the Input column: text box. Choose <u>OK</u>.

The Minitab Output

Inverse Cumulative Distribution Function

Normal with mean = 12.0000 and standard deviation = 2.0000ı

```
P( X <= x)          x
   0.9900      16.6527
```

Figure 6.10

The resulting Minitab output, as shown in Figure 6.10, in the Session window indicates that the largest 1% of the time to register distribution is made up of values that are greater than 16.6527 minutes. If students are disconnected fter 16.65 minutes, only 1% of all students registering would be disconnected prior to completing their registration.

Exercises

6.8 Determine each of the areas under the standard normal (z) curve.
   a. To the left of z = -1.72
   b. To the left of z = -0.65
   c. To the right of z = 1.15
   d. To the right of z = -2.52
   e. Between z = -1.65 and z = 1.28
   f. Between z = 0.56 and z = 1.34

6.9 Let z be a random variable having a normal distribution with $\mu = 0$ and $\sigma = 1$. Determine the value z^* to satisfy the following conditions.
   a. $P(z < z^*) = 0.95$

   Follow these steps to determine the value of z^*.

   (i) Determine the value of z^*.
   Choose **Calc**>**Probability Distributions**> **Normal.** Darken the **I**nverse cumulative probability: option button. Place 0 in the **M**ean: text box. Place 1 in the **S**tandard deviation: text box. Darken the Input co**n**stant: option button. Place 0.95 (the area to the left of z^*) in the Input co**l**umn: text box. Choose **OK**.

   b. $P(z < z^*) = 0.01$
   c. $P(z > z^*) = 0.05$
   d. $P(z > z^*) = 0.5$
   e. $P(z^* < z < z^*) = 0.90$
   f. $P(z^* < z < z^*) = 0.95$
   g. $P(z^* > z > z^*) = 0.02$

6.10 Cholesterol readings for males are thought to approximate a normal distribution with $\mu = 200$ and $\sigma = 25$. Find the percentage (proportion) of males having cholesterol readings
   a. greater than 230
   b. greater than 160
   c. less than 180
   d. less than 235

      e. between 160 and 220

      f. between 180 and 190

      g. between 225 and 250

      h. find the value of x^* such that 2% have cholesterol readings above x^*.

      i. find the value of x^* such that 20% have cholesterol readings below x^*.

6.11 The length of time to complete a placement examination is thought to approximate a normal distribution with $\mu = 160$ minutes and $\sigma = 20$ minutes. If sufficient time for only 85% of the individuals to complete the test, when should the test be terminated?

6.12 The heights of adult males are thought to approximate a normal distribution with $\mu = 68$ inches and $\sigma = 2.5$ inches. How high should a long passageway in a confined area (where every available space is critical) be constructed so that 95% of the males can pass through the passageway without having to bend?

6.5   Checking for Normality

New Minitab Commands

  1. **Calc>Calculator** - Does arithmetic using an algebraic expression.

      a. Normal scores (NSCOR) - Calculates normal scores, which can be used to produce normal probability plots. In this section, you will use this command in constructing a normal probability plot. You can produce normal probability plots directly by using **Stat > Basic Statistics > Normality Test**. You will also use this command in this section to produce normal probability plots.

      b. Log 10 (LOGT) - Calculates logarithims to base 10.

  2. **Stat > Basic Statistics > Normality Test** - Generates a normal probability plot. The grid on the graph resembles the grids found on normal probability paper. The vertical axis has a probability scale; the horizontal axis, a data scale. A least-squares line is fit to the plotted points and drawn on the plot for reference. The line forms an estimate of the cumulative distribution function for the population from which data are drawn. Numerical estimates of the population parameters, m and s, are displayed with the plot. In this section, you will use this command to construct a normal probability plot to examine the assumption that the dsitribution from which these observations were drawn is normally distributed.

      Some of the most frequently used statistical methods are valid only when a sample x_1, x_2, ...x_n has come from a population distribution that is at least approximately normally distributed. One way to determine if an assumption of population normality is plausible is to construct a normal probability plot. A substantial linear pattern in a normal probability plot suggests that population normality is plausible. On the other hand, a systematic departure from a straight-line pattern (such as curvature in the plot) casts doubt on the legitimacy of assuming a normal population distribution.

Chapter   6

The Problem
The following observations represent the a sample of 20 observations of the miles
per gallon (mpg) for a popular four-cylinder model engine in a new car.

25.79	25.85	25.60	25.51	24.21	25.01	24.22
25.75	26.27	25.75	25.34	24.95	25.48	25.34
25.70	25.27	25.73	24.88	26.14	26.29	

Follow these steps to construct a normal probability plot to examine the assumption that the mpg distribution from which these observations were drawn is normal.

1. Enter data.
   Enter the 20 observations of the mpg in column C1 in the Data window. Name
   column C1 as Mpg.

2. Determine normal scores.
   Choose **Calc>Calculator**. Place NormalScores in the Store result in variable:
   text box. Select Statistics from the Functions: drop down dialog box. Choose
   Normal scores from the Statistics functions, placing NSCOR(number) in the
   Expression: text box. Replace number in the NSCOR(number) in the expres-
   sion with Mpg. Choose **OK** .

3. Create the normal probability plot.
   Choose **Graph>Plot.**
   Place NormalScores in the Y Graph variables: text box. Place Mpg in the
   X Graph variables: text box. Choose **OK.**

The Minitab Output

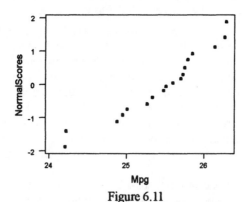

Figure 6.11

The linearity displayed in the normal probability plot shown in Figure 6.11 supports the assumption that the mpg distribution from which these observations were
drawn is normal.

Exercises

6.13 The value of equipment for equipment used by service personnel appear in the following table.

Follow these steps to construct a normal probability plot to examine the assumption that the value distribution from which these observations were drawn is normal.

40	76	34	48	67	67	65	56	55
96	44	43	35	39	38	37	30	27
33	80	39	61	44	45	59	39	

a. Enter data.
   Enter the observations of the values in column C1 in the Data window. Name column C1 as Values.

b. Construct a histogram.
   Choose **Stat>Basic Statistics>Display Descriptive Statistics**. Place Values in the Variables: text box. Choose **Graphs..**. Place a check in the Graphical summary checkbox. Choose **OK**. Choose **OK**. In this step, you have obtained a histogram of the data over which is drawn a normal curve.

c. Print the graphical summary.
   Choose **File>Print Graph.** After selecting the correct printer, choose **OK** to print the graphical summary.

   (i) Does the histogram suggest that the data support the assumption that the value distribution from which these observations were drawn is normal?

d. Create the normal probability plot.
   Choose **Stat >Basic Statistics > Normality Test**. Place Values in the Variables: text box. Place Value of Equipment in the Title: text box. Choose **OK**. In this step, you have used another method of obtaining a normal probability plot.

e. Print the normal probability plot.
   Choose **File>Print Graph.** After selecting the correct printer, choose **OK** to print the normal probability plot.

f. Observe that the points do not perfectly fit the straight line. Does the normal probability plot support the assumption that the value distribution from which these observations were drawn is normal?

g. Make a transformation.
   Choose **Calc>Calculator**. Place LogValues in the Store result in variable: text box. Select Logarithm from the Functions: drop down dialog box. Choose Log 10 from the Logarithmic functions, placing LOGT(number) in the Expression: text box. Replace number in the LOGT(number) in the expression with Values. Choose **OK**. In this step, you are transforming the data, attempting to determine if the transformed data will be normal.

h. Construct a histogram.
   Choose **Stat>Basic Statistics>Display Descriptive Statistics**. Place Log-

Values in the Variables: text box. Choose **Graphs..** Place a check in the Graphical summary checkbox. Choose **OK**. Choose **OK**. In this step, you have obtained a histogram of the LogValues over which is drawn a normal curve.

i. Print the graphical summary.

Choose **File>Print Graph.** After selecting the correct printer, choose **OK** to print the graphical summary.

j. Does the histogram of LogValues suggest that the the assumption that the distribution of LogValues is normal?

k. Create the normal probability plot.

Choose **Stat > Basic Statistics > Normality Test.** Place LogValues in the Variables: text box. Place Value of Equipment in the Title: text box. Choose **OK**. In this step, you have constructed a normal probability plot of the LogValues.

l. Print the normal probability plot.

Choose **File>Print Graph.** After selecting the correct printer, choose **OK** to print the normal probability plot.

m. Observe that the points again do not perfectly fit the straight line. Compare the two normal probability plots. Examine the extreme values of each plot. Does the transformation of the variable Values on the normal probability plot support the assumption that the distribution of LogValues is closer to a normal distribution?

6.14 Consider the following dollar values of collision damage to an automobile.

64	112	67	117	93
73	129	81	141	119
84	152	89	157	165
98	170	104	71	

Construct a normal probability plot, and comment on the appropriateness of a normal probability model.

6.15 The accompanying observations are DDT concentrations in the blood of 20 individuals.

24	26	30	35	35
38	39	40	40	41
42	42	52	56	58
61	75	79	88	102

Construct a normal probability plot, and comment on the appropriateness of a normal probability model.

6.16 The file a:scores.mtp contains data with regard to several variables. Construct a normal probability plot of the variable Math and comment on the appropriateness of a normal probability model.

6.6   Transformations

New Minitab Commands

1. **Calc>Calculator** - Peforms arithmetic using an algebraic expression. You can use arithmetic operations, comparison operations, logical operations, functions, and column operations. In this section, you will use this calculator to transform data using natural logarithms, square roots, and exponents.

    Many of the most frequently used statistical methods are valid only when the sample is selected at random from a population whose distribution is at least approximately normal. When the sample histogram shows a distinctly non-normal shape, it is common to use a transformation of the data to yield a distribution of transformed values which are more closely approximated by a normal curve.

    The square root transformation tends to pull in the long tail of the distribution on the right (make large data values smaller), but stretch it out on the left (make small data values larger). On the other hand, a logarithmic transformation tends to make small data values larger and large data values smaller.

The Problem - Exposure to Beryllium

(Example 7.20, Text) Exposure to beryllium is known to produce adverse effects on lungs as well as on other tissues and organs in both laboratory animals and humans. The paper "Time :Lapse Cinematographic Analysis of Beryllium - Lung Fibroblast Interactions" (*Envir. Reserach* (1983):34-43) reported the results of experiments designed to study the behavior of certain individual cells that had been exposed to beryllium. An important characteristic of such an individual cell is its interdivision time (IDT). IDT's were determined for a large number of cells both in exposed (treatment) and unexposed (control) conditons. The authors of the paper state, "The IDT distributions are seen to be skewed, but the natural logs do have an approximate normal distribution." Representative IDT data appear in the following table.

28.1	46.0	34.8	17.9	31.9	23.7	26.6
43.5	30.6	52.1	15.5	38.4	21.4	40.9
31.2	25.8	62.3	19.5	28.9	18.6	26.2
17.4	55.6	21.0	36.3	72.8	20.7	57.3
13.7	16.8	28.0	21.1	60.1	21.4	32.0
38.8	25.5	22.3	19.1	48.9		

Follow these steps to construct a histogram of IDT data and a histogram of transformed IDT data.

a. Enter the data.

   Enter the data in column C1. Name column C1 as IDT.

b. Transform the data.

   Choose **Calc>Calculator**. Place TransformedData in the Store result in variable: text box. Click on the down arrow for the Functions: drop down dialog box. Choose Logarithm. Choose Log 10. Double click on IDT to replace number with IDT in the LOGT function in the Expression: text box. Choose

OK.

c. Construct the frequency histogram for the IDT data.
Choose **Graph>Histogram...** Place IDT in the first row of the X Graph variables: text box. Choose **OK.**

d. Construct the frequency histogram for the transformed data.
Choose **Graph>Histogram...** Place TransformedData in the first row of the X Graph variables: text box. Choose **OK.**

The Minitab Output

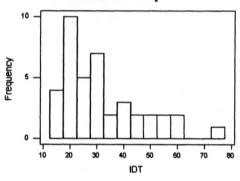

Figure 6.12

The graph window, as shown in Figure 6.12, indicates the general shape of the original IDT data.

The Minitab Output

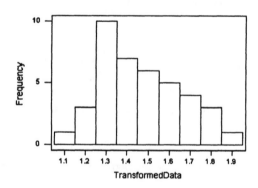

Figure 6.13

The graph window, as shown in Figure 6.13, indicates the general shape of the transformed IDT data.

Follow these steps to construct a normal probability plot of the IDT data and a normal probability plot of the transformed IDT data.

a. Create the normal probability plot for the IDT data.

Choose **Stat** >**Basic Statistics** > **Normality Test**. Place IDT in the Variables: text box. Place Normal Probability Plot of IDT in the Title: text box. Choose **OK**.

c. Create the normal probability plot for the transformed IDT data.

Choose **Stat** >**Basic Statistics** > **Normality Test**. Place TransformedData in the Variables: text box. Place Normal Probability Plot of Transformed IDT in the Title: text box. Choose **OK**.

The Minitab Output

Normal Probability Plot of IDT

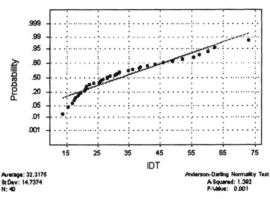

Figure 6.14

The lack of linearity displayed in the normal probability plot shown in Figure 6.14 supports the assertion that the original IDT data seem to be skewed, and is consistent with the histogram shown in Figure 6.12.

The Minitab Output

Normal Probability Plot of Transformed IDT

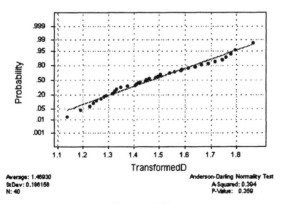

Figure 6.15

Chapter 6

The linearity displayed in the normal probability plot shown in Figure 6.15 supports the assertion that the $\log_{10}$ (IDT) transformation has an approximately normal distribution, and is consistent with the histogram shown in Figure 6.13.

Exercises

6.17 Ecologists have long been interested in factors that might explain how far north or south particular animal species are found. As part of one such study, the paper "Temperature and the Northern Distributions of Wintering Birds" (*Ecology* (1991):2274-2285) gave the accompanying body masses (in grams) for 50 different bird species that had previously been thought to have northern boundaries corresponding to a particular isotherm.

7.7	10.1	21.6	8.6	12.0	11.4	16.6
9.4	9.8	10.2	15.9	11.5	9.0	8.2
20.2	48.5	21.6	26.1	6.2	93.9	31.0
12.5	19.1	21.0	28.1	10.6	31.6	6.7
5.0	68.8	10.9	21.5	14.5	23.9	19.8
20.1	6.0	99.6	19.8	16.5	9.0	19.6
11.9	32.5	448.0	21.3	17.4	36.9	34.0
41.0						

a. Enter the data, naming column one as Grams. Construct a histogram of the data.

b. Introduce transformations of the data and construct histograms of the transformed data:

(i) Transform the data using logarithms and construct a histogram.

(ii) Transform the data using the transformation: $TransformedData = \frac{1}{\sqrt{Grams}}$ and construct a histogram of the transformed data.

Chapter 7
Sampling Variability
and Sampling Distributions

7.1 Overview

The inferential methods presented in subsequent chapters will use information contained in a sample to reach conclusions about one or more characteristics of the whole population. For example, let μ denote the true mean height of all students in all sections of your statistics class. To learn something about μ, a sample of n = 20 students might be selected and the height determined for each student in the sample. The sample data might produce a mean of $\bar{x} = 68.4$ inches. How close is this sample mean to the population mean, μ? If another sample of 20 were selected and $\bar{x}$ computed, would this second $\bar{x}$ value be near 68.4, or might it be quite different? If the sampling were repeated again, would there be variation in the sample means? Is there a distribution that describes the distribution of sample mean heights? Does the sampling distribution of sample means depend on the distribution of the original parent population from which the sample data were obtained?

These issues and others can be addressed by studying what is called the sampling distribution of $\bar{x}$. The sampling distribution of $\bar{x}$ provides information about the long-run behavior of $\bar{x}$. After reading this chapter you should be able to

1. Obtain Sample of Size n from a Parent Population (Simulation)
2. Construct a Histogram of the Distribution of Sample Means
3. Describe the Sampling Distribution of $\bar{x}$

7.2   Statistics and Sampling Variability

New Minitab Commands

1. **Calc>Random Data>Discrete** - Generates random data from a discrete distribution. In this section, you will enter a discrete distribution (table) into two columns of Minitab. You will then use this command, to generate random data, in conjunction with other commands in order to examine the sampling distribution of $\bar{x}$.

2. **Calc>Row Statistics** - Computes one value for each row in a set of columns. The statistic is calculated across the rows of the column(s) specified and the answers are stored in the corresponding rows of a new column. In this section, you will use this command to calculate row means.

3. **Calc>Random Data>Normal** - Generates random data from a normal distribution. In this section, you will enter the population mean and the population standard deviation of a normal distribution into appropriate text boxes to generate random data. You will use this command in conjunction with other commands in order to examine the sampling distribution of $\bar{x}$.

4. **Calc>Random Data>Binomial** - Generates random data from a binomial distribution. In this section, you will enter the number of trials and the probability of success into appropriate text boxes to generate random data. You will use this command in conjunction with other commands in order to examine the sampling distribution of $\bar{x}$.

Many investigations and research projects aim at drawing conclusions about how the values of some variable x are distributed in a population. Often, attention is focused on a single characteristic of that distribuion. Examples include:

1. $x =$ sodium content (milligrams) of a can of Campbell's chicken noodle soup, with interest centered on the mean sodium content μ of all cans of Campbell's chicken noodle soup

2. $x =$ amount of fill (ounces) of a bottle of soft drink, with interest focused on the variability of the amount of fill as described by σ, the standard deviation for the amount of fill population distribution

3. $x =$ ethnic background (a categorical variable) of a student at a particular university, with attention focused on π, the proportion of all students at the university of Hispanic background.

The usual way of obtaining information regarding the value of a population characteristic is by selecting a sample from the population. For example, to gain insight about the mean height of the height distribution of all students in all sections of your statistics class, we might select a sample of 20 students and determine their heights. The height of each student would be determined to yield a value of $x = height$. You could then construct a histogram of the 20 sample x values, and we could view this histogram as an approximation of the population distribution of x. In a similar fashion, we could view the sample mean, $\bar{x}$ as an approximation of μ, the mean of the population distribution. It would be nice if $\overline{x} = \mu$. It could be, but this is rarely the case: remember $\overline{x}$ is only an estimate of μ. An important question is: How good is this estimate?

Not only does the value of $\overline{x}$ for a particular sample from a population usually differ from μ, but the $\overline{x}$ values from different samples of 20 heights will usually result in different $\overline{x}$ values. It is this sample-to-sample variability that makes it so challenging to generalize from a sample to a population from which the sample was selected. To meet these challenges, we need to understand this sample to sample variability.

Quantities computed from values of the sample are called sample statistics. Just as a variable associates a value with every individual or object in a population and can be described by its distribution, a statistic associates a value with each individual sample in the population of samples. Thus, a statistic can also be described by a distribution. The distribution of a statistic is called its sampling distribution.

The Problem- Sampling from a Discrete Distribution

Consider a small population consisting of the board of directors of a day car center where the variable of interest is x = number of children for each board member.

Board member	Jay	Carol	Allison	Teresa	Lygia	Bob	Roxy	Kyle
Number of children	2	2	0	0	2	2	0	3

The population mean μ of the # of children is 1.38.

If the sampling is done with replacement there are 36 different possible outcomes when a sample of size 2 is selected from the board; these samples are shown in Table 7.1.. Observe that the example in the text illustrates sampling without replacement.

Sample	$\overline{x}$	Sample	$\overline{x}$	Sample	$\overline{x}$	Sample	$\overline{x}$
1. Jay Jay	2	10. Carol Allison	1	19. Allison Bob	1	28. Lygia Bob	2
2. Jay Carol	2	11. Carol Teresa	1	20. Allison Roxy	0	29. Lygia Roxy	1
3. Jay Allison	1	12. Carol Lygia	2	21. Allison Kyle	1.5	30. Lygia Kyle	2.5
4. Jay Teresa	1	13. Carol Bob	2	22. Teresa Teresa	0	31. Bob Bob	2
5. Jay Lygia	2	14. Carol Roxy	1	23. Teresa Lygia	1	32. Bob Roxy	1
6. Jay Bob	2	15. Carol Kyle	2.5	24. Teresa Bob	1	33. Bob Kyle	2.5
7. Jay Roxy	1	16. Allison Allison	0	25. Teresa Roxy	0	34. Roxy Roxy	0
8. Jay Kyle	2.5	17. Allison Teresa	0	26. Teresa Kyle	1.5	35. Roxy Kyle	1.5
9. Carol Carol	2	18. Allsion Lygia	1	27. Lygia Lygia	2	36. Kyle Kyle	3

Table 7.1

Table 7.1 gives the value of $\overline{x}$ for each of the 36 possible samples, and a histogram of these $\bar{x}$ values appears in Figure 7.1.

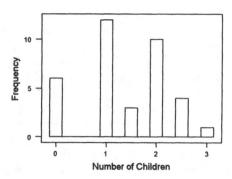

Histogram of Number of Children

Figure 7.1

Figure 7.1 indicates the true sampling distribution of the mean when the sample is of size 2 from this population. We see that the value $\bar{x} = 1$ is more common than any other value. However, $\bar{x}$ values can differ greatly from one sample to another, and also from the value of the population mean μ. If we were to select a random sample of size 2 (so that each different sample has the same chance of being selected) we could not count on the resulting value of $\bar{x}$ being very close to the population mean μ.

Let us approximate this sampling distribution using simulation. We can begin by selecting samples of size 2 from the population of the number of children and find the sample mean for each sample. A histogram can then be constructed to examine the sampling distribution of $\bar{x}$.

Follow these steps to simulate the sampling, with replacement, of size 2 of the # of children from the parent population of the number of children for each board member.

1. Enter Data.
   Enter the Number of children (0, 2, 3) in column C1. Enter the RelativeFrequecies (0.375, 0.5, 0.125) in column C2. Name column C1 as NumberChildren and column C2 as RelativeFrequency. In this step, you are constructing a discrete probability distribution based upon the number of children of the Board member's (3 of the 8 Board members have 0 children, implying $\frac{3}{8} = 0.375$, etc.)

2. Select samples.
   Choose **Calc>Random Data>Discrete.** Place 100 in the Generate rows of data: text box. Place Sample1 Sample2 in the Store in column(s): text box. Place NumberChildren in the Values in: text box. Place RelativeFrequency in the Probabilities in: text box. Choose **Ok.** In this step, you have selected a large number, 100, of random samples using a discrete probability distribution.

3. Calculate sample means.
   Choose **Calc>Row Statistics.** Darken the Mean Statistic option button. Place Sample1 Sample2 in the Input variables: text box. Place Means in the Store

result in: text box. Choose **OK**. In this step, the sample means for each of the 100 samples of size 2 has been calculated and stored in the column Means.

4. Construct a graphic.
Choose **Stat>Basic Statistics>Display Descriptive Statistics**. Place Means in the Variables: text box. Choose **Graphs..** Place a check in the Graphical summary checkbox. Choose **OK** . Choose **OK** .

The Minitab Output

Descriptive Statistics

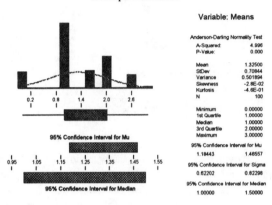

Variable: Means

Anderson-Darling Normality Test

| A-Squared: | 4.996 |
| P-Value: | 0.000 |

Mean	1.32500
StDev	0.70844
Variance	0.501894
Skewness	-2.8E-02
Kurtosis	-4.6E-01
N	100

Minimum	0.00000
1st Quartile	1.00000
Median	1.00000
3rd Quartile	2.00000
Maximum	3.00000

95% Confidence Interval for Mu

| 1.18443 | 1.46557 |

95% Confidence Interval for Sigma

| 0.62202 | 0.82298 |

95% Confidence Interval for Median

| 1.00000 | 1.50000 |

Figure 7.2

Figure 7.2 provides some insight into the behavior of the mean of the simulated random samples of size 2 from this population. A comparison of Figure 7.2 with Figure 7.1 indicates that the approximation to the true sampling distribution is reasonably close. Again, we can see that the value $\bar{x} = 1$ is more common than any other value.

Let us now select samples of size 3 from the population of the number of children and find the sample mean for each sample in order to examine the sampling distribution of $\bar{x}$.

Follow these steps to simulate the sampling of size 3 of the number of children from the parent population of the number of children for each board member.

1. Select samples.
Choose **Calc>Random Data>Discrete**. Place 100 in the Generate rows of data: text box. Place Sample3 in the Store in column(s): text box. Place NumberChildren in the Values in: text box. Place RelativeFrequency in the Probabilities in: text box. Choose **OK**.

2. Calculate sample means.
Choose **Calc>Row Statistics**. Darken the Mean Statistic option button. Place Sample1 Sample2 Sample3 in the Input variables: text box. Place Means in the Store result in: text box. Choose **OK**.

3. Construct a graphic.
Choose **Stat>Basic Statistics>Display Descriptive Statistics**. Place Means

in the Variables: text box. Choose **Graphs..** Place a check in the Graphical summary checkbox. Choose **OK** . Choose **OK** .

The Minitab Output

Descriptive Statistics

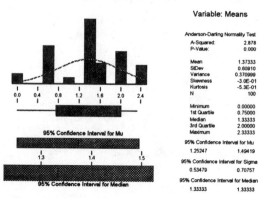

Variable: Means

Anderson-Darling Normality Test
A-Squared:	2.878
P-Value:	0.000
Mean	1.37333
StDev	0.60910
Variance	0.370999
Skewness	-3.0E-01
Kurtosis	-5.3E-01
N	100
Minimum	0.00000
1st Quartile	0.75000
Median	1.33333
3rd Quartile	2.00000
Maximum	2.33333

95% Confidence Interval for Mu
1.25247	1.49419

95% Confidence Interval for Sigma
0.53479	0.70757

95% Confidence Interval for Median
1.33333	1.33333

Figure 7.3

Figure 7.3 provides even more insight into the behavior of the distribution of sample means from this population. You can clearly see that the sampling distribution is becoming approximately normally distributed and that the standard deviation has become smaller. The density histogram of the statistic $\bar{x}$ can be regarded as the sampling distribution of the statistic $\bar{x}$. The sampling distribution of $\bar{x}$ provides some important information about the behavior of the statistic $\bar{x}$ and how it relates to μ.

The Problem- Sampling from a Normal Distribution

Consider a population of all students in all sections of your statistics class where the variable of interest is x = height. Let us select 100 random samples of size n = 20 from the parent population where the mean μ is 68.4 inches and the population standard deviation σ is 2.5 inches and approximate the sampling distribution of $\bar{x}$ using simulation.

Follow these steps to simulate the sampling of size n = 20 of the heights of students from the parent population of the heights of all students in all sections of your statistics class.

1. Select samples.

   Choose **Calc>Random Data>Normal.** Place 100 in the Generate rows of data: text box. Place C1-C20 in the Store in column(s): text box. Place 68.4 in the Mean: text box. Place 2.5 in the Standard deviation: text box. Choose **OK**. In this step, you have selected 100 random samples of size 20 from the parent population.

2. Calculate sample means.

   Choose **Calc>Row Statistics.** Darken the Mean Statistic option button. Place C1-C20 in the Input variables: text box. Place Means in the Store result in: text box. Choose **OK**. In this step, you have calculated the means of the 20 simulated heights.

3. Construct a graphic.

   Choose **Stat**>**Basic Statistics**>**Display Descriptive Statistics**. Place Means in the **V**ariables: text box. Choose **Graphs..** Place a check in the **G**raphical summary checkbox. Choose **OK** . Choose **OK** .

The Minitab Output

Descriptive Statistics

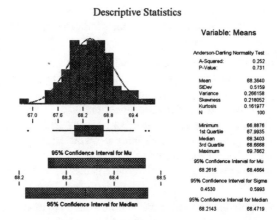

Figure 7.4

When you examine the data window you can see that the value of $\overline{x}$ (Row1, Row2, etc.) differs from one random sample to another (sampling variability). Some samples produced $\overline{x}$ values larger than $\mu = 68.4$ inches, whereas others produced $\overline{x}$ values smaller than μ. Figure 7.4 provides some evidence about the nature of the sampling distribution of $\overline{x}$: the mean of the distribution of $\overline{x}'s$ in this sampling distribution is fairly close to μ and the standard deviation is relatively small.

Exercises

7.1 Consider the "game" where you win $1.00 when two fair coins are tossed and you observe exactly one head - and - of course, you lose $1.00 when the fair coin is tossed and you observe exactly zero or two heads. This "game" satisfies the requirement for a binomial probability distribution where the number of trials is 2 and the probability of success (observing a head) is 0.5.

Follow these steps to simulate the sampling of size n = 20 (playing the game 20 times) from this parent population of outcomes.

   a. Select samples.

   Choose **C**alc>**R**andom Data>**B**inomial. Place 50 in the **G**enerate rows of data: text box. Place C1-C20 in the **S**tore in column(s): text box. Place 2 in the **N**umber of trials: text box. Place 0.5 in the **P**robability of success: text box. Choose **OK**.

   b. Calculate sample means.

   Choose **C**alc>**R**ow Statistics. Darken the **M**ean Statistic option button. Place C1-C20 in the Input **v**ariables: text box. Place Means in the **S**tore result in: text box. Choose <u>O</u>K.

c. Construct a graphic of the binomial experiment.
Choose <u>S</u>tat><u>B</u>asic Statistics><u>D</u>isplay Descriptive Statistics. Place C1 in the <u>V</u>ariables: text box. Choose **Graphs..** Place a check in the <u>G</u>raphical summary checkbox. Choose <u>O</u>K . Choose <u>O</u>K .

d. Construct a graphic of the distribution of sample means.
Choose <u>S</u>tat><u>B</u>asic Statistics><u>D</u>isplay Descriptive Statistics. Place Means in the <u>V</u>ariables: text box. Choose **Graphs..** Place a check in the <u>G</u>raphical summary checkbox. Choose <u>O</u>K . Choose <u>O</u>K .

e. In what ways are the two sampling distributions similar? In what ways are they different.

7.2 Consider the population: {1,2,3,4,5}. Calculate the summary statistics for this data set in the form of a graphical summary. Select 20 random samples of size 2, with replacement, from this population. Calculate the sample mean for each of the samples. Construct the sampling distribution of $\bar{x}$ in the form of a graphical summary. (Hint: there are a number of ways of dealing with this problem in Minitab. Examine **<u>C</u>alc><u>R</u>andom Data><u>D</u>iscrete, <u>C</u>alc><u>R</u>andom Data><u>S</u>ample From Columns** or
<u>C</u>alc><u>R</u>andom Data><u>I</u>nteger.) Do the $\bar{x}$ values differ a lot from sample to sample, or do they tend to be similar?

7.3 Consider a population consisting of the following five values, which represent the amount of money spent on a gift for a friend: {9, 12, 15, 10, 16}. Calculate the summary statistics for this data set in the form of a graphical summary. Select 20 random samples of size 2 from the population. Calculate the sample mean for each of the samples. Construct the sampling distribution of $\bar{x}$ in the form of a graphical summary. (Hint: there are a number of ways of dealing with this problem in Minitab. Examine **<u>C</u>alc><u>R</u>andom Data><u>D</u>iscrete**, or
<u>C</u>alc><u>R</u>andom Data><u>S</u>ample From Columns.) Do the $\bar{x}$ values differ a lot from sample to sample, or do they tend to be similar?

7.4 The uniform probability distribution is a rectangular distribution where the probability of a particular value occurring is the same for each value. Let the uniform probability distribution occur over the interval from 0 to 10. Place samples of size 50 in columns C1-C20. Calculate the summary statistics for this data set (select column C1) in the form of a graphical summary. Calculate the sample mean for each of the random samples of size 20 from this population. Construct the sampling distribution of $\bar{x}$ in the form of a graphical summary.

7.3   The Sampling Distribution of a Sample Mean

When the objective of a statistical investigation is to draw an inference about the population mean μ, it is natural to consider the sample mean $\bar{x}$. In order to understand how inferential procedures based on $\bar{x}$ work, we must first examine how sampling variability causes $\bar{x}$ to differ in value from one sample to another sample. The behavior of $\bar{x}$ is is described by its sampling distribution. The sample size n and characteristics of the population - its shape, mean value μ, and standard deviation σ - are important considerations in determining properties of the sampling distribution of $\bar{x}$.

The Problem- Sampling from a Normal Distribution

Analyses attempting to determine the impact of particular food groups on health assessments have considered the amounts of dietary fat and its relationship to cancer rates. The distributions of dietary fat may be approximately normal in shape. The apparent values for μ and σ were $\mu = 50$ grams and $\sigma = 17.4$ grams.

Follow these steps to select 500 samples of size n = 5, 10, 20 and 30 of the amount of dietary fat from the parent population.

1. Select samples.

    Choose **Calc>Random Data>Normal.** Place 500 in the Generate rows of data: text box. Place C1-C30 in the Store in column(s): text box. Place 50 in the Mean: text box. Place 17.4 in the Standard deviation: text box. Choose **OK**.

2. Calculate sample means.

    Choose **Calc>Row Statistics.** Darken the Mean Statistic option button. Place C1-C5 in the Input variables: text box. Place MeansFive in the Store result in: text box. Choose **OK**.

    Choose **Calc>Row Statistics.** Darken the Mean Statistic option button. Place C1-C10 in the Input variables: text box. Place MeansTen in the Store result in: text box. Choose **OK**.

    Choose **Calc>Row Statistics.** Darken the Mean Statistic option button. Place C1-C20 in the Input variables: text box. Place MeansTwenty in the Store result in: text box. Choose **OK**.

    Choose **Calc>Row Statistics.** Darken the Mean Statistic option button. Place C1-C30 in the Input variables: text box. Place MeansThirty in the Store result in: text box. Choose **OK**.

3. Construct graphics.

Construct the relative frequency histogram.

Choose **Graph**>**Histogram...** Place MeansFive in row 1 of the Graph variables: text box. Place MeansTen in row 2 of the Graph variables: text box.
Place MeansTwenty in row 3 of the Graph variables: text box. Place MeansThirty
in row 1 of the Graph variables: text box. Choose Options. Darken the Percent
option button for the Type of Histogram. Choose **OK**. Choose Frame>Multiple
Graphs. Darken the Each graph on a separate page option button for Generation of Multiple Graphs. Darken the Same X and same Y option button for
Scale of Graphs on Separate Pages. Choose **OK**. Choose **OK**.

The Minitab Output

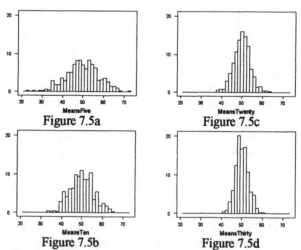

Figure 7.5a   Figure 7.5c

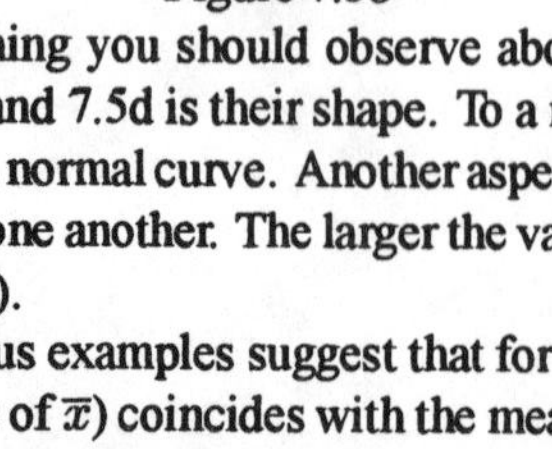

Figure 7.5b 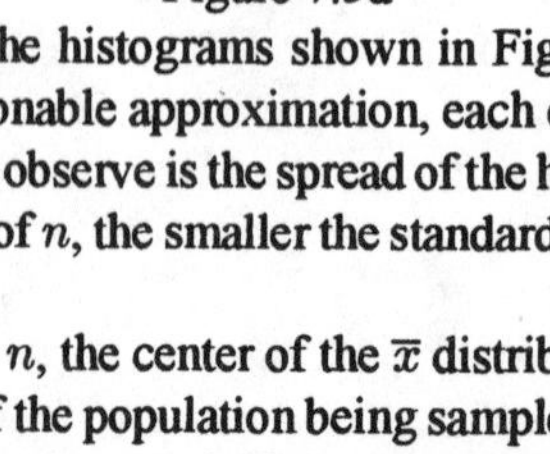   Figure 7.5d

The first thing you should observe about the histograms shown in Figures 7.5a,
7.5b, 7.5c and 7.5d is their shape. To a reasonable approximation, each of the four
looks like a normal curve. Another aspect to observe is the spread of the histograms
relative to one another. The larger the value of n, the smaller the standard deviation
(the spread).

The previous examples suggest that for any n, the center of the $\bar{x}$ distribution (the
mean value of $\bar{x}$) coincides with the mean of the population being sampled, but the
spread of the $\bar{x}$ distribution decreases as n increases. Additionally, the histograms
then indicate that the standard deviation is smaller for large n than for small n. The
sample histograms also suggest that the $\bar{x}$ distribution is approximately normal in

shape. These observations are stated more formally in the following general rules.

General Rules Concerning the $\bar{x}$ Sampling Distribution
Let $\bar{x}$ denote the mean of the observations in a
random sample of size n from a population
having mean μ and standard deviation σ. Denote the mean
value of the $\bar{x}$ distribution by $\mu_{\bar{x}}$ and the standard deviation by $\sigma_{\bar{x}}$.
Then the following rules hold.

Rule 1. $\mu_{\bar{x}} = \mu$

Rule 2. $\sigma_{\bar{x}} = \frac{\sigma}{\sqrt{n}}$ This rule is approximately correct as long as no more than 5% of the population is included in the sample.

Rule 3.   When the population distribution is normal, the sampling distribution of $\bar{x}$ is also normal for any sample size n.

Rule 4.   (Central Limit Theorem) When n is sufficiently large, the sampling distribution of $\bar{x}$ is well approximated by a normal curve, even when the population distribution is not itself normal.

Recall that a variable is standardized by subtracting the mean value and then dividing by its standard deviation.

If n is large or the population distribution is normal, the standardized variable

$$z = \frac{\bar{x} - \mu}{\sigma_{\bar{x}}} = \frac{\bar{x} - \mu}{\frac{\sigma}{\sqrt{n}}}$$

has (approximately) a standard normal (z) distribution.

The Problem- Amount of Fill

The amount of fill (fluid ounces) put into a soft drink container is claimed to be approximately normally distributed with $\mu = 12$ ounces and standard deviation $\sigma = 0.1$ ounce. Consider selecting $n = 16$ containers and determining the amount of fill of each container. Because x is approximately normally distributed, $\bar{x}$ is also approximately normally distributed. If the claim is correct, the $\bar{x}$ sampling distribution has a mean $\mu_{\bar{x}} = \mu_x = 12$ and a standard deviation of $\sigma_{\bar{x}} = \frac{\sigma_x}{\sqrt{n}} = \frac{0.1}{\sqrt{16}} = .025$. Let us calculate the probability that the mean amount of fill is between 11.95 ounces and 12.02 ounces.

Follow these steps to calculate the probability that the mean amount of fill is between 11.95 ounces and 12.02 ounces.

1. Enter data.

   Enter the values of 11.95 and 12.02 in column C1. Name column C1 as Ounces.

2. Determine the area to the left of the values.

   Choose **Calc>Probability Distributions> Normal.** Darken the Cumulative probability: option button. Place 12 in the Mean: text box. Place .025 in the Standard deviation: text box. Darken the Input column: option button. Place Ounces in the Input constant: text box. Choose **OK.**

Chapter   7

The Minitab Output

Cumulative Distribution Function

```
Normal with mean = 12.0000 and s.d. = 0.025

       x       P( X <= x)
   11.9500       0.0228
   12.0200       0.7881
```

Figure 7.6

Then the probability that the mean amount of fill is between 11.95 ounces and 12.02 ounces is found by subtracting the two areas under the normal curve: $0.7881 - .0228 = .7653$ ($P(11.95 < x < 12.02) = 0.7653$). If the x distribution is as described and the claim is correct, a sample average fluid content based on 16 observations will fall in the interval from 11.95 ounces to 12.02 ounces about 76.53% of the time.

The Problem- Paint Coverage

A manufacturer of paint for the stripes on the road claims that the average coverage of its white road paint is 460 sq ft/gal and that the standard deviation $\sigma = 0.8$ sq ft/gal. Let x denote the coverage of one randomly selected gallon of white road paint. The probability distribution of x is unknown. Consider selecting $n = 36$ gallons and determining the coverage of each one and let $\bar{x}$ denote the resulting sample average coverage. The sample size, $n = 36$, is sufficiently large to invoke the Central Limit Theorem and regard the $\bar{x}$ distribution as being approximately normal. The standard deviation of the $\bar{x}$ distribution is $\sigma_{\bar{x}} = \frac{\sigma_x}{\sqrt{n}} = \frac{0.8}{\sqrt{36}} = 0.13333$ sq ft/gal. If the manufacturer's claim is correct, we know that $\mu_{\bar{x}} = \mu_x = 460$ sq ft/gal. Suppose the sample resulted in a mean of $\bar{x} = 459.7$ sq ft/gal. Does this result indicate that the manufacturer's claim is incorrect?

Follow these steps to calculate the probability that the mean amount of coverage is significantly less than the manufacturer's claim.

1. Enter data.

   Enter the values of 459.7 in column C1. Name column C1 as Coverage.

2. Determine the area to the left of the value.

   Choose **Calc>Probability Distributions> Normal.** Darken the **C**umulative probability: option button. Place 460 in the **M**ean: text box. Place .013333 in the **S**tandard deviation: text box. Darken the Input column: option button. Place Coverage in the Input constant: text box. Choose **OK**.

The Minitab Output

Cumulative Distribution Function

Normal with mean = 460.000 and s.d. = 0.133330

x	P(X <= x)
459.7000	0.0122

Figure 7.7

The value $\bar{x} = 459.7$ does indicate that the mean amount of coverage is significantly less and casts doubt on the manufacturer's claim. Values of $\bar{x} = 459.7$ or less would be observed less than 1.22% of the time when a random sample of size 36 is taken from a population with a mean of 460 and a standard deviation 0.8.

Exercises

7.5 Five year-old children are thought to have heights that are approximately normally distributed with mean $\mu = 100$ cm and standard deviation $\sigma = 6$ cm. If a random sample of $n = 16$ children are selected and $\bar{x}$ denotes the sample mean height, what is the probability that the mean height is between 98 cm and 102 cm?

7.6 The time that one randomly selected individual waits to be seated at a restaurant has a uniform probability distribution (over the interval from 0 to 10 minutes) where the mean is 5 minutes and the standard deviation is 2.8868 minutes. If a random sample of $n = 16$ individuals are selected and $\bar{x}$ denotes the sample mean waiting time, what is the probability that the mean time exceeds 6 minutes? is less than 4.05 minutes?

7.7 "Time headway" in traffic flow is the elapsed time between the time that one car finishes passing a fixed point and the instant that the next car begins to pass that point. The times have an exponential distribution with mean $\mu = 6$ minutes and standard deviation $\sigma = 6$ minutes. If a random sample of $n = 25$ cars are selected, what is the probability that the mean time is between 4.5 minutes and 7.6 minutes? exceeds 8.4 minutes?

7.8 An airplane has a total baggage limit of 6000 lb. Let x = total weight of the baggage checked by an individual passenger, be approximately normally distributed with mean $\mu = 50$ pounds and standard deviation $\sigma = 20$ pounds. If 100 passengers board a flight, what is the probability that the total weight of their baggage will exceed the limit? (Hint: With $n = 100$, the total weight exceeds the limit precisely when the average weight $\bar{x}$ exceeds 6000/100.)

Chapter 8
Estimation Using a
Single Sample

8.1 Overview

The objective of inferential statistics is to use sample data to increase our knowledge about the corresponding population. Often, data is collected to obtain information that allows the investigator to estimate the value of some population characteristic, such as a population mean, μ, or a population proportion, π. This could be accomplished by using the sample data to arrive at a single number that represents a plausible value for the characteristic of interest. Alternatively, one could report an entire range of plausible values for the characteristic. These two estimation techniques, point estimation and interval estimation, are addressed in this chapter. After reading this chapter you should be able to

1. Make Point Estimates (Proportions, Means)
2. Construct Large-Sample Confidence Intervals (Proportions, Means)
3. Construct a Small-Sample Confidence Interval (Mean)

8.2 Point Estimation

The usual way of obtaining information regarding the value of a population characteristic, such as a population mean, μ, or a population proporiton, π, is by selecting a sample from the population. A *point estimate* of a population characteristic is a single number that is based on sample data and represents a plausible value of the characteristic. For example, a survey by a public interest group might report that 585 of 1000 individuals favor a proposal to lower the drunk-driving blood alcohol level from 0.10 to 0.08. The sample proportion, p, is a point estimate of π; in this case that is $p = \frac{585}{1000} = 0.585$. As a second example, sample data might suggest that 50 calories (from fat in a serving) is a plausible value for μ, the true mean calorie content (from fat in a serving) in Banana Nut Crunch cereal. This is the value stated on the package. In this example, 50 is a *point estimate* of μ.

The Problem - Cereals and Sodium

A recent study reported the results of a statistical experiment comparing the amounts of sodium in various categories of cereal. A random sample of the sodium contents of cereals are given in the following table.

220	200	120	300	0	10	240	253
293	230	345	187	360	0	200	213
210	53	22	0	280	187	227	267

Follow these steps to produce a summary of the data containing the numerical summaries.

1. Enter data.
   Enter the sodium content of the cereals in column C1. Name column C1 as Sodium.

2. Obtain a summary.
   Choose <u>S</u>tat><u>B</u>asic Statistics><u>D</u>isplay Descriptive Statistics. Place Sodium in the <u>V</u>ariables: text box. Choose <u>O</u>K .

The Minitab Output

Descriptive Statistics

Variable	N	Mean	Median	Tr Mean	StDev	SE Mean
Sodium	24	184.0	211.5	184.4	112.8	23.0

Variable	Min	Max	Q1	Q3
Sodium	0.0	360.0	69.7	263.5

Figure 8.1

Suppose that a point estimate of μ, the true mean sodium content, is desired. An obvious choice of a statistic for estimating μ is the sample mean $\overline{x}$. However, there are other possibilities. Figure 8.1 provides some of those other possibilities. The three estimates of the center of the population are $\overline{x} = \frac{\sum x}{n} = \frac{4417}{24} = 184.0$, the sample median $= \frac{210+213}{2} = 211.5$ and the 5% trimmed mean $= 184.4$. The estimates differ somewhat from each other. The choice from among them should depend on which statistic tends to produce an estimate closest to the true center of the distribution.

When the population distribution is normal, then $\overline{x}$ has a smaller standard deviation than any other unbiased statistic for estimating μ.

When the population distribution is symmetric with heavy tails compared to the normal curve, a trimmed mean is a better statistic for estimating μ.

Observe that in this example, the sample mean, $\overline{x} = 184.0$ and the trimmed mean, $TrMean = 184.4$ are actually quite close.

The sample variance $s^2 = \frac{\sum(x-\overline{x})^2}{n-1}$ is a good choice for obtaining a point estimate of the population variance σ^2. It can be shown that s^2 is an unbiased statistic for estimating σ^2; that is, whatever the value of σ^2, the sampling distribution of s^2 is centered on that value. It is precisely for this reason - to obtain an unbiased statistic - that the divisor $n - 1$ is used. Let σ^2 denote the true variance in sodium content in cereals. Using the sample variance, s^2 to provide a point estimate of σ^2 yields $s^2 = \frac{292908}{23} = 12735$. An obvious choice of a statistic for estimating the population standard deviation σ is the sample standard deviation s. For the data in this example, $s = \sqrt{12735} = 112.85$. Unfortunately, the fact that s^2 is an unbiased statistic does not imply that s is an unbiased estimate for estimating σ.

Exercises

8.1 A random sample of ten houses in a particular area, each of which is heated with natural gas, is selected, and the amount of gas (therms) used juring the month of January is determined for each house. The resulting observations are as follows.

103	156	118	89	125
147	122	109	138	99

a. Calculate three point estimates of the center of the distribution.

b. Calculate a point estimate for σ^2.

8.2 The following amounts represent the fees charged by a local delivery service for small items within the city.

4.24	3.76	3.28	6.40	5.98	3.37	3.08	3.11	4.00	3.22
4.54	4.10	4.97	4.23	4.87	5.47	3.06	3.16	5.32	4.95
5.76	4.11	7.18	5.19	4.87	3.82	4.04	4.04	3.96	7.20
4.39	4.21	4.04	6.01	8.23	4.88	5.17	6.00	3.00	4.07
5.32	5.55	4.26	5.78	4.93	4.11	4.24	4.32	3.17	4.28

a. Calculate three point estimates of the center of the distribution.

b. Calculate a point estimate for σ^2.

8.3 High temperatures (in degrees Fahrenheit) for August were recorded with the following results:

83	71	86	88	87	86	83	95	96	85
100	95	91	84	78	85	92	85	86	98
95	96	92	94	93	83	86	81	82	97
83									

a. Calculate three point estimates of the center of the distribution.

b. Calculate a point estimate for σ^2.

8.3   A Large-Sample Confidence Interval

New Minitab Commands

1. **Stat>Basic Statistics> 1 Proportion** -Performs perform a hypothesis test of the proportion and computes a confidence interval. In this section, you will calculate a confidence interval for the proportion of one or more variables.

2. **Stat>Basic Statistics>1-Sample-Z** - Performs a one-sample z-test or calculates a confidence interval for the mean. You need to know the population standard deviation to perform a z-test or calculate a z-confidence interval. In this section, you will darken the Confidence interval: option button to calculate a separate one-sample confidence interval for the mean of one or more variables.

You have just seen how to use a sample statistic to produce a point estimate of a population characteristic. The value of a point estimate depends on which sample is selected, and different samples usually yield different estimates, due to chance differences, of the population characteristic. Rarely is the point estimate from the sample exactly equal to the true value of the population characteristic. While a point estimate may represent the best single number guess for the value of the population characteristic, it is not the only plausible value.

Suppose, that instead of reporting a single point estimate as the single most credible value for the population characteristic, we report an interval of reasonable values based on the sample data. For example, we might be confident that the mean "-commuting distance", μ, of students is in the interval from 10.23 miles to 11.54 miles. The narrowness of this interval implies that we have rather precise information about the value of μ. If, with the same degree of confidence, we could state only that μ was in the interval from 0.5 miles to 20.3 miles, it would be clear that we had relatively imprecise knowledge of the value of μ.

> A confidence interval for a population characteristic is an interval of plausible values for the characteristic. It is constructed so that, with a chosen degree of confidence, the value of the characteristic will be captured indside the interval.
> The confidence level associated with a confidence interval estimate specifies the success rate of the method used to construct the interval.

The Problem - Confidence in the Economy

A survey of adults aged 18 to 75 addressed the question of public confidence in the economy. The level of confidence in the economy continuining to sustain the economic growth exhibited in the last year was of interest. Respondents were asked: "Do you expect the economy to continue to grow as it has in this last year?." The responses are contained in the file: economy.mtp.

Follow these steps to construct a 95% confidence interval for the population proportion, π, of respondents who responded yes:

1. Open the worksheet.
   Choose **File**>**Open Worksheet**. Select the file a:economy.mtp. Choose **OK**.

2. Construct the confidence interval.
   Choose **Stat**>**Basic Statistics**> **1 Proportion.** Darken the Samples in columns option button. Place Response in the Samples in columns: text box. Choose Options. Choose the default Level: of confidence 95.0. Accept the (default) value of 0.5 in the Test proportion: text box. Place a check in the Use test and interval based on normal distribution: checkbox. Choose **OK** . Choose **OK** .

The Minitab Output

```
Test and CI for One Proportion: Response

Test of p = 0.5 vs p not = 0.5

Success = Yes

Variable      X     N  Sample p       95.0% CI        Z-Value  P-Value
Response    802  1000  0.802000  (0.777302, 0.826698)   19.10    0.000
```

Figure 8.2

Based on the sample, plausible values of π, the true proportion of yes responses is in the interval from 0.777302 to 0.826698, as indicated in Figure 8.2. A 95% confidence level is associated with the method used to produce this interval estimate.

It is tempting to say that there is a 95% chance that π is between 0.777302 and 0.826698. The 95% refers to the percentage of all possible samples resulting in an interval that includes π. Said another way, if we take sample after sample form the population and use each one separately to compute a 95% confidence interval, in the long run approximately 95% of those intervals will capture π.

The Problem - Data Temperatures on the Internet
The temperatures from 530 stations throughout the U.S. for the September, 1996
are recorded in the file a:temps.mtp. (National Climatic Data Center - http://www.louisville.edu/groups/libr
www/ekstrom/govpubs/federal
/agencies/commerce/clim.html) Let μ denote the true mean temperature for September, 1996. Although σ, the true population standard deviation, is not known,
we will assume that $\sigma = 2.7$.
Follow these steps to construct a 95% confidence interval for the temperatures for
September, 1996:

1. Open the worksheet.
   Choose **File**>**Open Worksheet**. Select the file a:temps.mtp. Choose **OK**.

2. Construct the confidence interval.
   Choose **Stat**>**Basic Statistics**>**1-Sample-Z**. Place Temps in the **V**ariables:
   text box. Darken the **C**onfidence interval option button. Choose the default
   Level: of confidence 95.0. Place 2.7 in the **S**igma: text box. Choose **OK** .

The Minitab Output

Confidence Intervals

```
The assumed sigma = 2.70

Variable     N      Mean     StDev  SE Mean      95.0 % CI
Temps       530   61.180    2.686    0.117  ( 60.950,  61.410)
```

Figure 8.3

Based on the sample, plausible values of μ, the true mean temperature for September, 1996 is in the interval from 60.95°F to 61.41°F, as indicated in Figure 8.3.
A 95% confidence level is associated with the method used to produce this interval estimate.
It is tempting to say that there is a 95% chance that μ is between 60.95 and 61.41.
The 95% refers to the percentage of all possible samples resulting in an interval
that includes μ. Said another way, if we take sample after sample form the population and use each one separately to compute a 95% confidence interval, in the
long run approximately 95% of those intervals will capture μ.

Exercises

8.4 Each person in a random sample of 20 students at a particular university was
    asked whether he or she was registered to vote. The responses (R = registered,
    N = not registered) were as follows. Use this data to estimate π, the true pro-
    portion of all students at the university who were registered to vote, and place
    a 95% confidence on π.

    R   R   N   R   N   N   R   R   R   N
    N   R   R   R   R   R   N   R   R   R

    Write a brief statement interpreting the confidence interval.

8.5 The article "Ban on Smoking Shown to Improve Lung Health" (Los Angles Times, Dec 9, 1998) reported the results of a study of San Francisco bartenders. Of 39 bartendeers who reported respiratory problems prior to a ban on smoking in bars, 21 were symptom-free two months after the ban began. Suppose that it is reasonable to regard this group of 39 bartenders as a random sample of bartenders with reported respirator problems.

Follow these steps to construct a 95% confidence interval for the population proportion, π, of bartenders with respiratory problems who were sympton-free two months after the smoking ban.

   a. Construct the confidence interval.

   Choose **Stat**>**Basic Statistics**> **1 Proportion.** Darken the Summarized data option button. Place 39 in the Number of trials: text box. Place 21 in the Number of successes: text box. Choose Options. Choose the default Level: of confidence 95.0. Accept the (default) value of 0.5 in the Test proportion: text box. Place a check in the Use test and interval based on normal distribution: checkbox. Choose **OK** . Choose **OK** .

   Write a brief statement interpreting the confidence interval.

8.6 The file CaCrime contains the rates of violent crimes per 1,000 residents in 1994, as reported to the FBI by local authorities for 306 locations. Let μ denote the true mean population for those locations. Although σ, the true population standard deviation, is not known, we will assume that $\sigma = 38000$.

Follow these steps to construct a 90% confidence interval for the variable population(s) of California communities:

   a. Open the worksheet.

   Choose **File**>**Open Worksheet**. Select the file a:CaCrime.mtp. Choose **OK**.

   b. Construct the confidence interval.

   Choose **Stat**>**Basic Statistics**>**1-Sample-Z.** Place Population in the Variables: text box. Darken the Confidence interval option button. Place 90.0 in the Level: (of confidence) text box. Place 38000 in the Sigma: text box. Choose **OK** .

   c. Write a statement interpreting the 90% confidence interval.

8.7 Fifty high school students recently took exams designed to measure their knowledge in the areas of reading, writing, math, science and civics. These test scores are recorded in the file a:scores.mtp. Let μ denote the true mean math test score for all students taking this exam. Although σ, the true population standard deviation, is not known, we will assume that $\sigma = 7.8$.

   a. Construct a 90% confidence interval for the variable Math.

   b. Construct a 95% confidence interval for the variable Math.

   c. Construct a 98% confidence interval for the variable Math.

   d. As the level of confidence increases, describe what happens to the width of the confidence interval.

8.8 Place 90%, 95% and 98% confidence intervals on the variable Reading in the file a:scores.mtp.

a. Write a statement interpreting each confidence interval.

8.4   A Small-Sample Confidence Interval

New Minitab Commands

1. <u>Stat</u>><u>B</u>asic Statistics><u>1</u>-Sample-t - Performs a one sample t-test or t-confidence interval for the mean. In this section, you will darken the Confidence interval: option button to calculate a separate one-sample confidence interval for the mean of one or more variables when the population standard deviation is not known.

The large-sample confidence interval for μ is appropriate whatever the shape of the population distribution. This is because it is based on the Central Limit Theorem, which states that when n is sufficiently large, the $\overline{x}$ sampling distribution is approximately normal for any population distribution. When n is small, the Central Limit Theorem does not apply. one way to proceed in the small-sample case is to make a specific assumption about the shape of the population distribution and then use an interval that is valid under that assumption.

If the sample size, n, is small then the shape of the $\overline{x}$ sampling distribution may not be approximately normal. However, when the population distribution itself is normal, the $\overline{x}$ sampling distribution is approximately normal even for small sample sizes. Since σ is usually unknown, we must estimate σ^2 with the sample variance s^2, resulting in the standardized variable

$$t = \frac{\overline{x} - \mu}{\frac{s}{\sqrt{n}}}$$

The Problem - Blood Pressure

Actual blood pressure values for 12 randomly selected individuals are as follows:

117.4	108.1	121.8	127.2	129.8	113.5
133.0	131.2	108.5	127.6	121.9	113.5

A normal probability plot of this data appears in Figure 8.

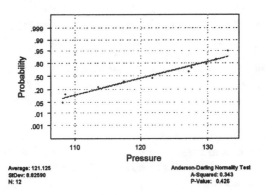

Normal Probability Plot

Average: 121.125
StDev: 8.82590
N: 12

Anderson-Darling Normality Test
A-Squared: 0.343
P-Value: 0.426

Figure 8.4

The plot, as shown in Figure 8.4, is reasonably straight, so it seems plausible that the population distribution is approximately normal.

Follow these steps to construct a 95% confidence interval for the blood pressure values:

1. Enter Data.

   Enter the blood pressure values in column C1. Name column C1 as Pressure.

2. Construct the confidence interval.

   Choose <u>S</u>tat><u>B</u>asic Statistics><u>1</u>-Sample-t. Place Pressures in the <u>V</u>ariables: text box. Darken the <u>C</u>confidence interval option button. Choose the default <u>L</u>evel: of confidence 95.0. Choose <u>OK</u>.

The Minitab Output

Confidence Intervals

```
Variable      N      Mean    StDev  SE Mean      95.0 % CI
Pressure     12    121.13     8.83     2.55  ( 115.52,  126.73)
```

Figure 8.5

With 95% confidence, the population mean blood pressure is estimated to be between 115.52 and 126.73, as indicated in Figure 8.5. Remember that the 95% confidence level implies that if the same formula is used to calculate intervals for sample after sample randomly selected from the population, in the long run 95% of the intervals will capture μ between the lower and upper confidence limits.

Exercises

8.9 An anthropologist measured the heights of a random sample of individuals from an isolated tribe and found the following heights:

70.84	73.45	71.18	72.20	71.86
72.20	73.22	70.95	71.97	71.96
71.52	73.67	73.22	72.20	71.97

Use the given data to construct a 90% confidence interval for the true mean height of the individuals in the tribe. Write a brief statement interpreting the confidence interval.

8.10 To assess the effects of a diet restricted in fat, saturated fat, and cholesterol 27 individuals participated in a dietary intervention study. The weight loss (for men) at the end of the ad libitum low-fat phase of the study were recorded with the following values (negative values indicate weight loss; positive values indicate weight gain:

4.6	7.9	-4.0	5.7	0.0	-3.1	-9.4	-5.8	-2.7
-0.2	-5.2	-5.9	-2.5	-3.2	-1.1	-3.6	0.2	0.2
-6.2	-6.9	-0.1	-5.1	5.8	-1.7	1.4	-10.5	-3.4

Use the given data to construct a 95% confidence interval for the true mean weight loss of the individuals in the study. Write a brief statement interpreting the confidence interval.

8.11 The ages of students in two sections of a statistics course were included in a survey conducted at the start of class. The data is included in the file a:ages.mtp. Use the given data to construct a 98% confidence interval for the true mean age of the students. Write a brief statement interpreting the confidence interval.

8.12 The sodium content of a number of different "breakfast toaster bars" is provided in the following table:

190	210	220	200	205	195	215	225	205	210
185	225	195	210	190	190	225	200	220	195
190	240	225	212	210	195	230	230	218	215
200	205	205	220	220	210	240	210	225	225
240	215	210	240	240	230	210	215	235	240

Construct a 99% confidence interval for the true mean sodium content. Write a brief statement interpreting the confidence interval.

8.13 The following amounts represent the fees charged by a local delivery service for small items within the city.

4.24	3.76	3.28	6.40	5.98	3.37	3.08	3.11	4.00	3.22
4.54	4.10	4.97	4.23	4.87	5.47	3.06	3.16	5.32	4.95
5.76	4.11	7.18	5.19	4.87	3.82	4.04	4.04	3.96	7.20
4.39	4.21	4.04	6.01	8.23	4.88	5.17	6.00	3.00	4.07
5.32	5.55	4.26	5.78	4.93	4.11	4.24	4.32	3.17	4.28

Construct a 90% confidence interval for the true mean fee. Write a brief statement interpreting the confidence interval.

Chapter 9
Hypothesis Testing
Using a Single Sample

9.1 Overview

In the previous chapter, we considered situtations in which the primary goal was to estimate the unknown value of some population characteristic. Sample data may also be used to decide if some claim or *hypothesis* about a population characteristic is plauible. This chapter addresses the issue of analyzing sample data to determine if a *hypothesis* about a population characteristic is plausible. Hypothesis testing methods presented in this chapter can be used to determine whether the sample data provides strong support for rejecting or failing to reject a *hypothesis*. After reading this chapter you should be able to

1. Perform a Large-Sample Hypothesis Test
    a. For a Proportion
2. Perform a Hypothesis Test
    a. For a Proportion
    a. For a Population Mean
3. Examine Errors in Hypothesis Testing

Under most conditions it is impossible or unrealistic to study an entire population to obtain the value of the population characteristic of interest. Statisticians have developed techniques that enable us to draw inferences about population parameters from sample statistics. This particular statistical decision making-tool is hypothesis testing. Hypothesis tests are used to investigate theories concerning population characteristics. Minitab may be used to make inferences about the value of a population parameter.

9.2 Hypotheses and Test Procedures

A hypothesis is a claim or statement either about the value of a single population characteristic or about the values of several population characteristics. The following are examples of legitimate hypotheses:

$\mu = 1000$ where μ is the mean number of characters in an email message

$\pi < .01$ where π is the proportion of email messages that are undeliverable.

A criminal trial is a familiar situation in which a choice between two competing claims must be made. In the U.S., the person accused of the crime is initially presumed to be innocent. Only strong evidence to the contrary will cause the presumption of innocence to be rejected in favor of a guilty verdict.

A test of hypotheses is a method for using sample data to decide between two competing claims (hypotheses) about a population characteristic. As in a U.S. judicial proceeding, we shall initially assume that a particular hypothesis, called the *null hypothesis* and designated as H_0, is the correct one. We then consider the evidence (the sample data), and we only reject the null hypothesis in favor of the competing hypothesis, called the *alternative hypothesis* and designated as H_a, if there is *convincing* evidence against the null hypothesis.

9.3   Errors in Hypothesis Testing

Once hypotheses have been formulated, we need to make a method fo using sample data to determine whether H_0 should be rejected. The decision rule that we use for this purpose is called a test procedure. Just as a jury trial may reach the wrong verdict in a trial, there is some chance that the use of a test procedure on sampling data may lead us to the wrong conclusion.

One erroneous conclusion in a criminal trial is for a jury to convict an innocent person, and another is for a guilty person to be set free. Similarly, there are two different types of errors that might be made when making a decision in a hypothesis-testing problem. One type of error involves rejecting H_0 even though H_0 is true. The second type of error results from failing to reject H_0 when it is false.

> Type I error   the error of rejecting H_0 when H_0 is true
> Type II error   the error of failing to reject H_0 when H_0 is false

No reasonable test procedure comes with a guarantee that neither type of error will be made; this is the price paid for basing an inference on a sample. With any procedure, there is some chance that a Type I error will be made, and there is also some chance that a Type II error will result.

9.4   Large Sample Tests for a Population Proportion

New Minitab Commands

1. <u>S</u>tat><u>B</u>asic Statistics> 1 <u>P</u>roportion -Performs perform a hypothesis test of the proportion and computes a confidence interval. In this section, you will perform a hypothesis test of the proportion.

   Now that some general concepts of hypothesis testing have been introduced, we are ready to turn our attention to the development of procedures for using sample information to choose between the null and alternative hypotheses. There are two possibilities - we will either reject H_0 or we will fail to reject H_0. The fundamental concept behind hypothesis testing is this: We reject the null hypothesis if the observed sample is very unlikely to have occurred when H_0 is true. In this section, we consider testing hypotheses about π, the population proportion.

Perhaps the most common inference of all is an inference concerning a proportion. Generally, we will let π represent the proportion of individuals or objects in a specified population that possess a certain property. A random sample of n individuals or objects is to be selected from the population. the sample proportion

$$p = \frac{x = \text{number that possess property}}{n}$$

is the natural statistic for making inferences about π.

When the sample proportion is to be tested against a hypothesized population proportion, we will use the test statistic

$$z = \frac{p' - p}{\sqrt{\frac{pq}{n}}} \text{ where } p' = \frac{x}{n}$$

The Problem - Domestic Violence

Recent research has focused on the offender's presence - on the police decision to arrest or not arrest the offender when responding to violent domestic incidents. Previous research has indicated that approximately 48% of offenders were absent when the police arrived. Suppose that 102 offenders were absent out of 210 violent domestic incidents. Does this data support the theory that the proportion of absent offenders is different from 48%, at the .05 level of significance? The hypotheses to be tested are

$$H_0 : \pi = .48$$
$$H_a : \pi \neq .48$$

The null hypothesis will be rejected only if there is convincing evidence that $\pi \neq$ .48 (that is, strong evidence against H_0.

Follow these steps to perform the hypothesis test.

1. Perform the hypothesis test.

    Choose <u>S</u>tat>**B**asic Statistics> **1 P**roportion. Darken the Summarized <u>d</u>ata option button. Place 210 in the Number of <u>t</u>rials: text box. Place 102 in the Number of <u>s</u>uccesses: text box. Choose O<u>p</u>tions. Accept the default <u>L</u>evel: of confidence 95.0. Enter 0.48 in the <u>T</u>est proportion: text box. Accept the default values of not equal from the <u>A</u>lternative drop down list box. Place a check in the <u>U</u>se test and interval based on normal distribution: checkbox. Choose **OK** . Choose **OK** .

The Minitab Output

Test and CI for One Proportion

Test of p = 0.48 vs p not = 0.48

Sample	X	N	Sample p	95.0% CI	Z-Value	P-Value
1	102	210	0.485714	(0.418117, 0.553312)	0.17	0.868

Figure 9.1

The Session window displays the results of the population proportion test, as shown in Figure 9.1. The Minitab output indicates an observed Z value of 0.17, and the p-value is 0.868 (implying $p > .05$). Since p is greater than α, the null hypothesis $H_0 : \pi = .48$ is not rejected. We thus do not have sufficient evidence to suggest that the true proportion is significantly different than 48%. There is not sufficient evidence to conclude that the proportion of absent offenders is different from 48%.

The Problem - A Placebo Effect

A placebo is a medication that looks real but contains no medically active ingredients. The placebo effect describes the phenomenon of improvement in the condition of a patient taking the placebo, when the patient has been taking no medically active ingredients. A manufactured cream, advertized as being able to reduce wrinkles and improve the skin, was compared to a placebo. Physicians placed the placebo on one side of the face and the advertized cream on the other side of the face. The number of responses indicating an improvement to the side of the face where the placebo was administered is recorded. Use the information contained in the file placebo.mtb to test the placebo effect. Assume that the placebo is ineffective; the probability of improvement is 0.5, at the .05 level of significance. The hypotheses to be tested are

$$H_0 : \pi = 0.5$$
$$H_a : \pi \neq 0.5$$

The null hypothesis will be rejected only if there is convincing evidence that $\pi \neq 0.5$ (that is, strong evidence against H_0.

Follow these steps to perform the hypothesis test.

1. Open the worksheet.
   Choose **File>Open Worksheet**. Select the file a:placebo.mtp. Choose **OK**.

2. Perform the hypothesis test.
   Choose **Stat>Basic Statistics> 1 Proportion**. Darken the Samples in columns: option button. Place Improvement in the Samples in columns: text box. Choose Options. Accept the default Level: of confidence 95.0. Enter 0.5 in the Test proportion: text box. Accept the default values of not equal from the Alternative drop down list box. Place a check in the Use test and interval based on normal distribution: checkbox. Choose **OK**. Choose **OK**.

The Minitab Output

Test and CI for One Proportion: Improvement

Test of p = 0.5 vs p not = 0.5

Success = Yes

Variable	X	N	Sample p	95.0% CI	Z-Value	P-Value
Improvement	484	1000	0.484000	(0.453026, 0.514974)	-1.01	0.312

Figure 9.2

The Session window displays the results of the population proportion test, as shown in Figure 9.2. The Minitab output indicates an observed Z value of -1.01, and the p-value is 0.312 (implying $p > .05$). Since p is greater than α, the null hypothesis $H_0 : \pi = 0.5$ is not rejected. We thus do not have sufficient evidence to suggest that the true proportion is significantly different than 50%. There is not sufficient evidence to conclude that the proportion of patients showing improvement is different from 50%.

Exercises

9.1 In a survey of 300 white pine trees, 51 were found to be infected with a tree borer. Does this data support the theory that the proportion of infected tree is greater than 15%, at the .05 level of significance?
   Follow these steps to perform the hypothesis test.

a. Perform the hypothesis test.

Choose **Stat**>**Basic Statistics**> **1 Proportion.** Darken the Summarized data option button. Place300 in the Number of trials: text box. Place 51 in the Number of successes: text box. Choose Options. Accept the default Level: of confidence 95.0. Enter 0.15 in the Test proportion: text box. Select greater than from the Alternative drop down list box. Place a check in the Use test and interval based on normal distribution: checkbox. Choose **OK** . Choose **OK** .

9.2 At a popular restaurant, 20% of the customers have placed an order from the desert selection in the past. After employing a new baker, the employees are instructed to ask customers if they wish to order desert whenever an order is placed, with the objective being to determine if the percentage of customers ordering desert has changed from historical 20% with the employment of a new baker. Out of the 432 customers served, 109 order desert. Use Minitab to test the hypothesis at the .01 level of significance.

a. Use the Minitab output to organize the hypothesis test in the traditional format (via the recommended steps).

9.3 Drug testing of job applicants is becoming increasing common. The Associated Press (May 24, 1990) reported that 12.1% of those tested in California tested positive. Suppose that this figure has been based on a sample of size 600, with 73 testing positive. Does this sample support a claim that more than 10% of job applicants in California test positive for drug use? Use Minitab to test the hypothesis at the .01 level of significance.

a. Use the Minitab output to organize the hypothesis test in the traditional format (via the recommended steps).

9.4 Seat belts help prevent injuries in automobile accidents, but they certainly don't offer complete protection in extreme situtations. A sample of 319 front-seat oc-cupants involved in head-on collisions in a certain region resulted in 95 who sustained no injuries ("Influencing Factors on the Injury Severity of Restrained Front Seat Occupants in Car-to-Car Head-on Collisions", *Accid.Anal. and Prev.* (1995):143-150). Does this suggest that the true (population) propor-tion of uninjured occupants exceeds .25? Use Minitab to test the hypothesis at the .01 level of significance.

a. Use the Minitab output to organize the hypothesis test in the traditional format (via the recommended steps).

9.5 A survey of television viewers in a local area indicated that 48 of 600 polled were viewing a program broadcast by the local station. Does this data support the theory that the proportion of viewers watching the program is less than 10%, at the .05 level of significance? Use Minitab to test the hypothesis at the .01 level of significance.

a. Use the Minitab output to organize the hypothesis test in the traditional format (via the recommended steps).

9.5   Hypothesis Tests for a Population Mean

New Minitab Commands

1. <u>S</u>tat><u>B</u>asic Statistics><u>1</u>-Sample t - Performs a one sample t-test or t-confidence interval for the mean. In this section, you will use this command to perform one sample t-tests.

    The procedures for testing hypotheses about a population mean μ are based on the same results that led to the confidence intervals in Chapter 8.Since it is rarely the case that σ, the population standard deviation, is known, we will focus our attention on the procedure for the procedure where σ is assumed to be unknown. Here we shall restrict consideration to the case where the original parent is assumed to be a normal population distribution.

    When x_1, x_2, ...x_n constitute a random sample of size n from a normal distribution, the probability distribution of the standardized variable is

$$t = \frac{\bar{x} - \mu}{\frac{s}{\sqrt{n}}}$$

the t distribution with $n-1$ degrees of freedom.

In most situations the population standard deviation, σ, is unknown. Minitab can use Student's t test to make inferences about the value of the population parameter μ when σ is unknown. This procedure may be applied to samples of all sizes where the assumption is that the parent population is approximately normally distributed.

The Problem - "Supermoms"

Mothers having full-time careers and also acting as the full-time homemaker mother have been a recent phenomena. One recent study focused on the total number of (a)hours of work that mothers having full-time careers spent working and (b) hours spent on caring for their homes. Mothers who worked at a paid job of 35 or more hours per week and had at least one child at home were asked to respond to a questionaire. That questionaire provided the following observations with regard to the number of hours spent working and also caring for their home.

$$73.1 \quad 65.1 \quad 45.8 \quad 60.0 \quad 69.3 \quad 75.9 \quad 77.6 \quad 73.6$$
$$66.6 \quad 88.4 \quad 83.3 \quad 86.3 \quad 66.0 \quad 56.1 \quad 71.6$$

Assuming that the hours are normally distributed, does the data suggest that the population mean number of hours spent caring for their home is different from 50 hours at the .05 level of significance?

Follow these steps to perform the hypothesis test.

1. Enter data.

    Enter the observations into column C1. Name column C1 as Hours.

2. Calculate the test statistic.

    Choose <u>S</u>tat><u>B</u>asic Statistics><u>1</u>-Sample t. Place Hours in the <u>V</u>ariables: text box. Darken the option button for <u>T</u>est mean:. Place 40 in the <u>T</u>est mean: text box. Choose the option of greater than in the <u>A</u>lternative: drop down dialog box. Choose G<u>r</u>aphs. Place a check in the <u>H</u>istogram of data checkbox. Choose <u>O</u>K. Choose <u>O</u>K.

The Minitab Output

Histogram of Hours

(with Ho and 95% t-confidence interval for the mean)

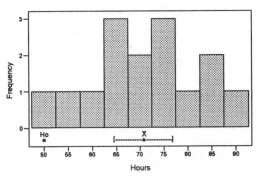

Figure 9.8

Minitab produces a histogram of the data consistent with the summary statistics. The histogram indicates the position of the value of $\mu = 50$ from the null hypothesis as well as a 95% confidence interval for the mean, using the sample data. Observe that the $H_0 : \mu = 50$ does not overlap the confidence interval.

The Minitab Output

T-Test of the Mean
Test of mu = 50.00 vs mu > 50.00

Variable	N	Mean	StDev	SE Mean	T	P
Hours	15	70.78	10.97	2.83	7.34	0.0000

Figure 9.9

Minitab produces a number of descriptive statistics (N, Mean, StDev, SE Mean), as well as the observed t value of 7.34, and the p-value of 0.0000 (implying $p < 0.001$). Since p is less than α, the null hypothesis $H_0 : \mu = 50$ is rejected. We thus do have sufficient evidence to suggest that the true mean number of hours spent working and also caring for their home is significantly greater than 50 hours. The evidence actually suggests that employed mothers worked the equivalent of nearly two full-time jobs.

Exercises

9.6 Many consumers pay careful attention to stated nutritional contents on packaged foods when making purchases. It is therefore important that the information on packages be accurate. A random sample of $n = 12$ frozen dinners of a certain type was selected from production during a particular period, and the calorie content of each one was determined. (This determination entails destroying the product, so a census would certainly not be desirable!) The stated calorie content is 240. The resuling observations follow.

255	244	239	242	265	245
259	248	225	226	251	233

Is the assumption of normality reasonable for this data? Is it reasonable to test hypotheses about the true average calorie content μ by using a t test? Does the data suggest that the population mean number of calories is different from 240 at the .05 level of significance?

Follow these steps to answer the above questions.

a. Enter data.

Enter the observations into column C1. Name column C1 as Calories.

b. Construct a boxplot.

Choose **Graph**>**Boxplot**. Place Calories in the Graph variables: Y (measurement) vs X (category) text box. Choose **OK**.

c. Construct a normal probability plot.

Choose **Stat**>**Basic Statistics**>**Normality Test**. Place Calories in the Variable: text box. Choose **OK**.

d. Calculate the test statistic.

Choose **Stat**>**Basic Statistics**>**1-Sample t**. Place Calories in the Variables: text box. Darken the option button for Test mean:. Place 240 in the Test mean: text box. Choose the (default) option of not equal in the Alternative: drop down dialog box. Choose Graphs. Place a check in the Histogram of data checkbox. Choose **OK**. Choose **OK**.

9.7 A recent article addressed discussed the amount of heat generated by certain types of electronic componets and the various heat disipaters. A sample of 16 components produced the following observations.

3.949	3.928	4.877	4.334	4.077	4.335	4.171	4.538
4.163	4.332	4.328	4.656	4.462	4.584	4.730	4.446

Is the assumption of normality reasonable for this data? Is it reasonable to test hypotheses about the true average heat μ by using a t test? Do these observations indicate that the amount of heat produced is different from the theoretical value of 4.5, at the .05 level of significance?

9.8 A study in biochemical embryology examined the chemical content of freshly laid, unfertilized eggs. The content of 12 eggs provided the following concentrations of lead in the eggs.

.6043	.6174	.6397	.6143	.6277	.6598
.6228	.6341	.6045	.5778	.6192	.6212

Is the assumption of normality reasonable for this data? Is it reasonable to test hypotheses about the true average lead content μ by using a t test? Do these observations indicate that the amount of lead concentration is different from from the historical value of .63 ppm, at the .05 level of significance?

9.9 Size AA batteries are supposed to have voltages of 1.5 volts. A random sample of 15 AA batteries provided the following voltages.

| 1.563 | 1.678 | 1.487 | 1.453 | 1.377 | 1.524 | 1.554 | 1.501 |
| 1.453 | 1.386 | 1.607 | 1.605 | 1.278 | 1.385 | 1.349 | |

Is the assumption of normality reasonable for this data? Is it reasonable to test hypotheses about the true average voltage μ by using a t test? Do these observations indicate that the voltage is less than from from the required 1.5 volts, at the .05 level of significance?

9.6   Type I and Type II Error Probabilities

New Minitab Commands

1. **Calc>Row Statistics** - Computes one value for each row in a set of columns. The statistic is calculated across the rows of the column(s) specified and the answers are stored in the corresponding rows of a new column. In this section, you will use this command to calculate row means.

2. **Manip>Copy Columns** - Copies data from columns in the current worksheet to new columns, including all rows or a specified subset. In this section, you will use this command to copy extreme Z scores to a new column.

3. **Calc>Column Statistics** - Calculates various statistics on the column you select, displaying the results and optionally storing them in a constant. In this section, you will use this command to obtain particular information about one variable in the Data window.

Consider a machine that produces ball bearings. Because of variation in the machining process, bearings produced by this machine do not have identical diameters. Let μ denote the true average diamter for bearings currently being produced. Suppose that the machine was initially calibrated to achieve the design sprcification $\mu = .5$ in. However, the manufacturer is now concerned that the diameters no longeer conform to this specification. That is, the hypothesis $\mu \neq .5$ in. must now be considered a possibility. If sample evidence suggests that $\mu \neq .5$ in., the production process will be halted while recalibration takes place. Because this is costly, the manufacturer wants to be quite sure that $\mu \neq .5$ in. before undertaking recalibration.

Hypotheses and Alpha

Under these circumstances, the null hypothesis, $H_0 : \mu = .5$ will be tested against the alternative hypothesis, $H_a : \mu \neq .5$. The probability of rejecting the null hypothesis, H_0, when the null hypothesis is true is referred to as a Type I error. The probability of a Type I error is denoted as α ($P(TypeI) = \alpha$). Let us decide to let $\alpha = .05$. (The null hypothesis will then be rejected if $|Z| > 1.96$.)

Alpha and Chance

Let us consider this last example in more detail. To examine the relationship between α and pure chance, let us assume that a sample of 16 ball bearings are chosen at random from a large population and their diameters are measured. The population of diameters are assumed to be approximately normally distributed with $\mu = .5$ in. and $\sigma = .003$ in. We will test the hypothesis $H_0 : \mu = .5$ against the alternative $H_a : \mu \neq .5$, using $\alpha = .05$.

Follow these steps to simulate the results of running this test 20 times.

1. Enter random data.

   Choose **Calc**>**Random Data**>**Normal**. Place 20 in the Generate rows of data text box. Place C1-C16 in the Store in column(s): text box. Place 0.5 in the Mean: text box and .003 in the Standard deviation: text box. Choose **OK**.

2. Calculate the row mean(s).

   Choose **Calc**>**Row Statistics**. Darken the Mean option button. Place C1-C16 in the Input variables: text box. Place Mean in the Store result in: text box. Choose **OK**.

3. Calculate the test statistic(s).

   Choose **Calc**>**Standardize.** Place Mean in the Input column(s): text box. Place Zscore in the Store results in: text box. Darken the Subtract _____ and divide by _____ option button. Place 0.5 in the Subtract _____ text box and 0.00075 ($\frac{.003}{\sqrt{16}} = 0.00075$) in the divide by _____ text box. Choose **OK**.

4. Count the number of z-scores greater than 1.96 or less than -1.96.

   Choose **Manip**>**Copy Columns**. Place ZScore in the Copy from columns: text box. Place Code in the To columns: text box. Select the Use Rows option button. Darken the Use rows with numeric column option button. Place ZScore in the Use rows with numeric column text box. Place -500:-1.96 1.96:500 in the Use rows with column equal to (eg, -4.5, -2:3 14): text box. Choose **Ok**. Choose **OK**.

   Choose **Calc**>**Column Statistics**. Darken the N total option button. Place Code in the Input variable: text box. Choose **OK**. Look at the Session window to see the count of the total number of observation in Code.

The Minitab Data Window

C17	C18	C19
Mean	ZScore	Code
0.499728	-0.35420	-2.12416
0.499183	-1.25314	
...	...	
0.498654	-2.12416	
...	...	

Figure 9.10

In how many tests did you reject H_0 ? That is, how many times did you make the "incorrect decision"? In this simulation, the incorrect decision was made 1 out of 20 times or 5% of the time. In other words, a Type I error was committed 1 of the 20 times the test was conducted. Identify the zscore of each test. Are they all the same? Suppose $\alpha = .10$ was used? Does this change any of your decisions to reject or not reject the null hypothesis?

Beta and Chance

Beta is the probability of failing to reject the null hypothesis when the null hypothesis is false. If the null hypotheses is $H_0 : \mu = 0.5$ vs. the alternative $Ha : \mu \neq 0.5$, using $\alpha = .05$, then the null hypothesis would be rejected if

$|Z| > 1.96$. That would imply that the null hypothesis will be rejected for means greater than 245.8085 or less than 244.1915.

Follow these steps to simulate the results of running this test 20 times.

1. Enter random data.

   Choose **Calc>Random Data>Normal**. Place 20 in the Generate rows of data text box. Place C1-C16 in the Store in column(s): text box. Place .502 (a value greater than $0.5 + 1.96 \cdot \left(\frac{.003}{\sqrt{16}}\right)$!) in the Mean: text box and .003 in the Standard deviation: text box. Choose **OK**.

2. Calculate the row mean(s).

   Choose **Calc>Row Statistics**. Darken the Mean option button. Place C1-C16 in the Input variables: text box. Place Mean in the Store result in: text box. Choose **OK**.

3. Calculate the test statistic(s).

   Choose **Calc>Standardize**. Place Mean in the Input column(s): text box. Place Zscore in the Store results in: text box. Darken the Subtract _____ and divide by _____ option button. Place 0.5 in the Subtract _____ text box and 0.00075 ($\frac{.003}{\sqrt{16}} = 0.00075$) in the divide by _____ text box. Choose **OK**.

4. Count the number of z-scores in the interval from -1.96 to +1.96.

   Choose **Manip>Copy Columns**. Place ZScore in the Copy from columns: text box. Place Code in the To columns: text box. Select the Use Rows option button. Darken the Use rows with numeric column option button. Place ZScore in the Use rows with numeric column text box. Place -1.96:1.96 in the Use rows with column equal to (eg, -4.5, -2:3 14): text box. Choose **OK**. Choose **OK**. Choose **Calc>Column Statistics**. Darken the N total option button. Place Code in the Input variable: text box. Place Count in the Store result in: text box. Choose **OK**.

The Minitab Data Window

C17	C18	C19
Mean	ZScore	Code
0.502565	3.41956	1.84573
0.501384	1.84573	1.28121
...	...	.
0.501914	2.55239	
...	...	

Figure 9.11

In how many tests did you to fail to reject H_0 ? That is, how many times did you make the "incorrect decision"? In this simulation, the incorrect decision was made 8 out of 20 times or 40% of the time. In other words, an incorrect decision was made 8 out of 20 times and thus a Type II error was committed. How would this example change if a mean further away from 0.5 were chosen?

Chapter   9

Exercises

9.10 Use Minitab to illustrate Type I and Type II error rates for the hypotheses
$$H_0 : \mu = 8$$
$$H_a : \mu < 8$$
when sampling from a normal population with a known standard deviation of 3.4641. In this situation, the appropriate test statistic is
$$Z = \frac{\bar{x} - 8}{\frac{3.4641}{\sqrt{n}}}$$
Suppose we wish to use a .05 significance level for out test. We thus should reject the null hypothesis whenever $Z < -1.645$.

a. Use Minitab to generate 20 random samples of size 16 from a normal distribution with a mean of 8 and a standard deviation of 3.4641. Calculate the sample mean and Z for each sample. Count the number of samples which yield values of $Z < -1.645$, i.e. which reject the null hypothesis. Comment on your results.

b. Use Minitab to illustrate Type II error rates by generating 20 random samples of size 16 from a normal distribution with a mean of 6 and a standard deviation of 3.4641. Calculate the sample mean and Z for each sample. Count the number of samples which yield values of $Z > -1.645$, i.e. which reject the null hypothesis. Comment on your results.

9.11 Pizza Hut, after test-marketing a new product called the Bigfoot Pizza, concluded that introduction of the Bigfoot nationwide would increase its sales by more than 14% (*USA Today*, April 2, 1993). This conclusion was based on recording sales information for a sample of Pizza Hut restaurants selected for the marketing trial. With μ denoting the mean percent increase in sales for all Pizza Hut restaurants, consider using sample data to decide between
$$H_0 : \mu = 14$$
$$H_a : \mu > 14$$

a. Use Minitab to illustrate Type I error rates by generating 500 random samples of size 9 from a normal distribution with a mean of 14 and a standard deviation of 4.5. Calculate the sample mean and Z for each sample. Count the number of samples which yield values of $Z > 1.645$, i.e. which reject the null hypothesis. Comment on your results.

b. Use Minitab to illustrate Type II error rates by generating 500 random samples of size 9 from a normal distribution with a mean of 18 and a standard deviation of 4.5. Calculate the sample mean and Z for each sample. Count the number of samples which yield values of $Z < 1.645$, i.e. which reject the null hypothesis. Comment on your results.

c. Restart Minitab, then rework parts a and b changing the sample size to 16. What effect does changing the sample size have on Type I and Type II error rates?

9.12 Water samples are taken from water used for cooling as it is being discharged from a power plant into a river. It has been determined that as long as the

124

mean temperature of the discharged water is at most $150^\circ F$, there will be no negative effects on the river's ecosystem. To investigate whether the plant is in compliance with regulations that prohibit a mean discharge water temperature above $150^\circ F$, 50 water samples are taken at randomly selected times and the temperatures of each sample recorded. The resulting data will be used to test the hypotheses

$$H_0 : \mu = 150$$
$$H_a : \mu > 150$$

a. Use Minitab to illustrate Type I error rates by generating 20 random samples of size 50 from a normal distribution with a mean of 150 and a standard deviation of 10. (This would imply 20 rows and 50 columns.) Calculate the sample mean and Z for each sample. Count the number of samples which yield values of $Z > 1.645$, i.e. which reject the null hypothesis. Comment on your results.

b. Use Minitab to illustrate Type II error rates by generating 20 random samples of size 50 from a normal distribution with a mean of 160 and a standard deviation of 10. Calculate the sample mean and Z for each sample. Count the number of samples which yield values of $Z < 1.645$, i.e. which reject the null hypothesis. Comment on your results.

c. Restart Minitab, then rework part b changing the sample mean to 180. What effect does making the sample mean further away from the hypothesized value of 150 on the Type II error rate?

Chapter 10
Comparing Two Populations Or Treatments

10.1 Overview

Many investigations are carried out for the purpose of comparing two populations. For example, one study focused on the question of whether there are differences in consumer perceptions of retail price reductions when there is an extremely low sale price and when there is a moderately low sale price. The study included subjects who were presented with ads containing an unusually deep (about 50%) discount from the expected retail price and subjects who were presented with ads containing a discount of about 25% from the product's expected retail price. After exposure to the ads, both groups (of subjects) reported on their perceptions of the value of the advertised deal, resulting in data that led the researchers to conclude that there were significant differences between consumer's responses to the ads.

To reach this conclusion, hypothesis tests that compare the means of two different populations were used. This chapter addresses hypothesis tests and confidence intervals that can be used when comparing two populations or treatments on the basis of means. After reading this chapter you should be able to

1. Perform a Small-Sample Hypothesis Test
   Concerning the Difference Between
   Two Normal Population Means for Independent Samples
2. Obtain a Confidence Interval on the Difference Between
   Two Normal Population Means
3. Perform a Small-Sample Hypothesis Test
   Concerning the Difference Between
   Two Population Means Using Paired Samples
4. Obtain a Confidence Interval on the Difference Between
   Two Population Means for Paired Samples
5. Perform a Large-Sample Hypothesis Test
   Concerning the Difference Between
   Two Normal Population Proportions
6. Perform the Mann-Whitney Rank-Sum Test

Under most conditions it is impossible or unrealistic to study an entire population to obtain the value of the population characteristic of interest. Statisticians have developed techniques that enable us to draw inferences about population parameters from sample statistics. This particular statistical decision making-tool is hypothesis testing. Hypothesis tests are used to investigate theories concerning population characteristics. Minitab may be used to make inferences about the value of a population parameter.

10.2   Independent Samples

New Minitab Commands

1. **Stat>Basic Statistics>2-Sample t** - Performs an independent two-sample t-test and generates a confidence interval.

   a. **Samples in one column** - Choose if the groups are stacked in the same column, differentiated by subscript values (group codes) in a second column. In this section, you will use this command to perform a two-sample t-test where all the data is in one column, with the subscripts in a second column.

   b. **Samples in different columns** - Choose if the groups are in two separate columns. In this section, you will use this command to perform a two-sample t-test where the data is in two seperate columns.

Hypothesis Tests on $\mu_1 - \mu_2$

An investigator who wishes to compare two populations is often interested either in estimating the difference between two population means or in testing hypotheses about the difference between the two population means. In the small sample case, the test procedure that is appropriate requires the assumption that the two population distributions are normal. Normal probability plots can be used to check the plausibility of the normality assumptions.

When the two samples are independently selected from normal population distributions, the standardized variable

$$t = \frac{(\overline{x}_1 - \overline{x}_2) - (\mu_1 - \mu_2)}{\sqrt{\frac{s_1^2}{n_1} + \frac{s_2^2}{n_2}}}$$

has approximately a t distribution with

$$df = \frac{(V_1 + V_2)^2}{\frac{V_1^2}{n_1 - 1} + \frac{V_2^2}{n_2 - 1}} \text{ where } V_1 = \frac{s_1^2}{n_1} \text{ and } V_2 = \frac{s_2^2}{n_2}$$

df should be truncated (rounded down) to an integer.

The Problem - Tennis Elbow

Tennis elbow is thought to be aggravated by the impact experienced when hitting the ball. The paper "Forces on the Hand in the Tennis One-Handed Backhand" (*Int. J. of Sport Biomechanics* (1991):282-292) reported the force (N) on the hand just after impact on a one-handed backhand drive for six advanced players and for intermediate players. The following data is consistent with the research study.

Advanced	58.17	36.10	37.90	33.60	48.95	27.05	
Intermediate	37.31	13.11	24.21	10.51	25.11	19.31	18.11	23.51

Determine, at the .01 level of significance, whether the mean force after impact is greater for advanced tennis players than it is for intermediate players.

Follow these steps to perform the hypothesis test.

1. Enter data.

   Enter the observations for the Advanced players into column C1 and continue to enter the observations for the Intermediate players into column C1 directly underneath the observations for the Advanced players. Name column C1 as Force. Code the Advanced players as 1 by placing six 1's in column C2. Code the Intermediate players as 2 by placing eight 2's directly underneath the six 1's in column C2. Name column C2 as Players.

2. Calculate the test statistic.

Choose **Stat**>**Basic Statistics**>**2-Sample t**. Darken the option button for Samples in one column. Place Force in the Samples: text box. Place Players in the Subscripts: text box. Choose the option of greater than in the Alternative: drop down dialog box. Choose **OK**.

The Minitab Output

Two Sample T-Test and Confidence Interval

Twosample T for Force

Players	N	Mean	StDev	SE Mean
1	6	40.3	11.3	4.6
2	8	21.40	8.30	2.9

95% C.I. for mu (1) - mu (2): (6.3, 31.5)
T-Test mu (1) = mu (2) (vs >): T= 3.46 P=0.0043 DF= 8

Figure 10.1

Minitab produces a number of descriptive statistics (N, Mean, StDev, SE Mean), as well as the observed t value of 3.46, and the one-tailed p-value of 0.0043 (meaning $p < \alpha = .01$). Since p is less than α, the null hypothesis $H_0 : \mu_1 - \mu_2 > 0$ is rejected. We thus do have sufficient evidence to suggest that the true mean forces after impact are greater for Advanced players when compared to Intermediate players.

You will observe that the hypothesis test also produced a 95% confidence interval on $\mu_1 - \mu_2$.

The Problem - Sunbathing

Individuals who desire to look good and feel good about themselves often sunbathe in spite of the long-term risk of skin cancer. An experiment explored the attitude of a group of males and females towards taking a hypothetical new drug, with different levels of risk, that would induce a "suntan". The following data represent a sample of observations from the experiment.

"Attitude scores"

Females	43.3	38.8	40.2	41.3	46.1	41.1	42.2	44.1	40.4
Males	41.7	34.9	40.1	37.9	37	38.4	35.8		

Determine, at the .05 level of significance, whether there are differences in the "Attitude scores" between Females and Males.

Follow these steps to perform the hypothesis test.

1. Enter data.

Enter the observations for the Females into column C1. Name column C1 as Females. Enter the observations for the Males into column C2. Name column C2 as Males.

2. Calculate the test statistic.

Choose Stat>Basic Statistics>2-Sample t. Darken the option button for Samples in different columns. Place Females in the First: text box. Place Males in the Second: text box. Choose the (default) option of not equal in the Alternative: drop down dialog box. Choose OK.

The Minitab Output

Two Sample T-Test and Confidence Interval

Twosample T for Females vs Males

	N	Mean	StDev	SE Mean
Females	9	41.94	2.24	0.75
Males	7	37.97	2.37	0.90

95% C.I. for mu Females - mu Males: (1.43, 6.52)
T-Test mu Females = mu Males (vs not =): T= 3.40 P=0.0052 DF= 12

Figure 10.2

Minitab again produces a number of descriptive statistics (N, Mean, StDev, SE Mean), as well as the observed t value of 3.40, and the two-tailed p-value of 0.0052 (meaning p<α = .05). Since p is less than α, the null hypothesis $H_0 : \mu_1 - \mu_2 \neq 0$ is rejected. We thus do have sufficient evidence to suggest that the true mean "Attitude scores" are different for Females when compared to Males.

Again, you will observe that the hypothesis test also produced a 95% confidence interval on $\mu_1 - \mu_2$.

Confidence Intervals on $\mu_1 - \mu_2$

A small sample confidence interval for $\mu_1 - \mu_2$ can easily be obtained by using Minitab. The two-sample t confidence interval for $\mu_1 - \mu_2$ is

$$\bar{x}_1 - \bar{x}_2 \pm (t \text{ critical value}) \sqrt{\frac{s_1^2}{n_1} + \frac{s_2^2}{n_2}}$$

The critical t values is based on

$$df = \frac{(V_1 + V_2)^2}{\frac{V_1^2}{n_1-1} + \frac{V_2^2}{n_2-1}} \text{ where } V_1 = \frac{s_1^2}{n_1} \text{ and } V_2 = \frac{s_2^2}{n_2}$$

df should be truncated (rounded down) ton an integer. This confidence interval is valid when both population distributions are normal.

The Problem - Soil pH

Nine observations of surface-soil pH were made at each of two different locations at the Central Soil Salinity Research institute experimental farm, and the resulting data appeared in the article "Sodium-Calcium Exchange Equilibria in Soils as Affected by Calcium Carbonate and Organic Matter" (Soil Sci. (1984):109). Construct a 90% confidence interval for the difference in mean pH levels between the two locations. Interpret the resulting interval.

Site					pH				
A	8.53	8.52	8.01	7.99	7.93	7.89	7.85	7.82	7.80
B	7.85	7.73	7.58	7.40	7.35	7.30	7.27	7.27	7.23

Construct a 90% confidence interval for the difference in mean pH levels between the two locations. Interpret the resulting interval.

Follow these steps to construct the 90% confidence interval.

1. Enter data.

   Enter the observations for Site A into column C1. Name column C1 as Site A. Enter the observations for Site B into column C2. Name column C2 as Site B.

2. Calculate the confidence interval.

   Choose **Stat**>**Basic Statistics**>**2-Sample t**. Darken the option button for Samples in different columns. Place Site A in the First: text box. Place Site B in the Second: text box. Place 0.90 in the Confidence level: text box. Choose **OK**.

The Minitab Output

Two Sample T-Test and Confidence Interval

Twosample T for Site A vs Site B

	N	Mean	StDev	SE Mean
Site A	9	8.038	0.285	0.095
Site B	9	7.442	0.224	0.075

90% C.I. for mu Site A - mu Site B: (0.384, 0.808)
T-Test mu Site A = mu Site B (vs not =): T= 4.92 P=0.0002 DF= 15

Figure 10.3

Minitab again produces a number of descriptive statistics (N, Mean, StDev, SE Mean). With 90% confidence, the difference in mean pH levels between the two locations is estimated to be between 0.384 and 0.808. The 90% confidence interval implies that if the same formula is used to calculate intervals for sample after sample randomly selected from the population, in the long run 90% of these intervals will capture $\mu_1 - \mu_2$ between the lower and upper confidence limits.

Exercises

10.1 A required placement exam is given to all students as they enter a certain college. One group (Group 1) of students had a graphing calculator intensive math course in high school, while the second group (Group 2) had a traditional approach without the presence of a graphing calculator. A sample of the placement scores are below.

Group 1	25	29	27	21	36	14	28	21	22	31
Group 2	17	22	18	27	19	24	26			

Does the data provide sufficient evidence to indicate that those students who had a graphing calculator intensive math course had scores that differ significantly from those students who had the traditional approach? Use $\alpha = .05$.

10.2 The paper "Smoking during Pregnancy and Lactation and its Effects on Breast-Milk Volume" (*American J. of Clinical Nutrition* (1991):1011-1016) indicated that nonsmoking mothers had a significantly greater breast-milk volume than did smoking mothers. Data is shown below.

Smokers	621	793	593	545	753	655	895	767	714	598
Nonsmokers	947	945	1083	1202	973	981	930	745	903	899

Does the data support the stated conclusion? Test the relevant hypotheses using $\alpha = .01$.

10.3 Two filling machines are designed to place 8 ounces of raisins in each package. the following weights (in ounces) are taken for each machine.

Machine1	8.30	8.56	9.01	8.60	7.33	7.76	7.18	7.96
	7.08	6.68	6.84	7.75	8.02	8.72		

Machine2	7.80	8.33	5.06	7.85	7.83	8.95	8.02	8.90
	6.53	9.90	8.45					

Does the data provide sufficient evidence to indicate that the mean amount of fill differs significantly between the two machines? Use $\alpha = .05$.

10.4 Rainfall specimens from various sites in New Zealand were analyzed to determine the amount of sulphur in the rainwater. The sites were classified according to whether they were closer to the east or the west coast of the islands.

Eastern Sites:	.26	.13	.62	.40	.28	.23	.80	.32	.08	.09
	.19	.21	.58	.17	.61					

Western Sites:	1.15	1.20	.43	.46	.44	.25	.43	.43	.25	.43
	.83	.11	.60	.43	.23	.30	.22	.08	.07	.28

Use a 90% confidence interval to estimate the difference in the mean amount of sulphur in rainwater between the Eastern and Western sites.

10.3 Paired Samples

New Minitab Commands

1. **Stat>Basic Statistics>Paired t** - Performs a paired t-test. This is appropriate for testing the difference between two means when the data are paired and the paired differences follow a normal distribution. Use the Paired t command to compute a confidence interval and perform a hypothesis test of the difference between population means when observations are paired. A paired t-procedure matches responses that are dependent or related in a pairwise manner. This matching allows you to account for variability between the pairs usually resulting in a smaller error term, thus increasing the sensitivity of the hypothesis test or confidence interval. In this section, you will use this command to perform a paired t-test.

Hypothesis Tests on μ_d

Two samples are said to be *independent* if the selection of the individuals or objects that make up one of the samples has no bearing on the selection of those in the other sample. In some situations, an experiment with independent random samples is not necessarily the best way to obtain information concerning any possible difference between the populations. For example, suppose an investigator wants to determine whether a particular drug will significantly lower systolic blood pressure. A random sample of people would be given a placebo and a second random sample of people would be given the actual drug. The two samples are selected independently of one another. The researcher then uses the two-sample t test to conclude that there are no significant differences between the two groups. It is known that diet and body weight influence systolic blood pressure. Might it not be the case that the individual differences in body weight are hiding any effects due to the drug?

A more reasonable approach would be to match individuals by body weight. The researcher would find pairs of subjects so that the individual given the placebo and the individual given the actual drug in each pair were similar in weight. The factor weight could then be ruled out as a possible explanation for the lack of an observed difference between the two groups. Matching the subjects by weight results in two samples for which each observation in the first sample is coupled in a meaningful way with a particular observation in the second sample. Such samples are said to be **paired**. Paired samples often provide more information than would independent samples, because extraneous effects are screened out.

When sample observations from the first population are paired in some meaningful way with the sample observations from the second population, inferences can be based on the differences between the two observations between each sampled pair. The n sample differences can then be regarded as having been selected from a large population of differences. Let μ_d = mean value of the difference population and σ_d = standard deviation of the difference population. the relationship between μ_d and the two individual population means is $\mu_d = \mu_1 - \mu_2$. Therefore, when the samples are paired, inferences about $\mu_1 - \mu_2$ are equivalent to inferences about μ_d. Since inference about μ_d can be based on the n observed sample differences, the original two-sample problem becomes a familiar one-sample problem. When the two samples are paired and it is reasonable to assume that the population of differences is normal, the standardized variable

$$t = \frac{\bar{d} - \mu_d}{\frac{s_d}{\sqrt{n}}}$$

has approximately a t distribution with

$$df = n - 1.$$

The t confidence interval for μ_d is

$$\bar{x}_d \pm (t \text{ critical value}) \frac{s_d}{\sqrt{n}}$$

The Problem - Home Appraisals

A particular mortgage institution provides determines the maximum amount of mortgage money available for the purchase of a home in the following manner. Two appraisals are made independently by real estate appraisers appointed by the mortgage institution. The maximum amount of mortgage money is the average of the two appraisals. the following random sample of 12 homes were appraised:

					Home	
	1	2	3	4	5	6
Appraiser1	80.7	95.2	108.0	101.3	93.6	70.9
Appraiser2	80.4	94.7	105.2	101.6	93.2	70.2

	7	8	9	10	11	12
Appraiser1	100.8	107.1	95.1	100.5	103.4	115.6
Appraiser2	99.8	104.5	94.9	98.7	103.8	114.5

Does the data suggest that there are significant differences in the appraised values of the homes between the two appraisers, at the .05 level of significance?

Follow these steps to perform the hypothesis test.

1. Enter data.

   Enter the observations for Appraiser1 into column C1. Name column C1 as Appraiser1. Enter the observations for Appraiser2 into column C2. Name column C2 as Appraiser2. Be sure to enter the observations as matched pairs.

2. Calculate the test statistic.

   Choose <u>Stat</u>><u>Basic Statistics</u>><u>Paired t</u>. . Place Appraiser1 in the <u>First</u> sample: text box. Place Appraiser2 in the <u>Second</u> sample: text box. Choose <u>Options</u>. Accept the (default) level of 95.0 in the <u>Confidence</u> level: text box. Accept the (default) level of 0.0 in the <u>Test</u> mean: text box. Choose the (default) option of not equal in the <u>Alternative</u>: drop down dialog box. Choose **OK**.

The Minitab Output

Paired T-Test and CI: Appraiser1, Appraiser2

Paired T for Appraiser1 - Appraiser2

	N	Mean	StDev	SE Mean
Appraiser1	12	97.68	12.12	3.50
Appraiser2	12	96.79	11.78	3.40
Difference	12	0.892	1.035	0.299

95% CI for mean difference: (0.234, 1.549)
T-Test of mean difference = 0 (vs not = 0): T-Value = 2.98   P-Value = 0.012

Figure 10.4

Minitab again produces a number of descriptive statistics (N, Mean, StDev, SE Mean), as well as the observed t value of 2.98, and the two-tailed p-value of 0.012 (meaning $p < \alpha = .05$). Since p is less than α, the null hypothesis $H_0 : \mu_d \neq 0$ is rejected. We thus do have sufficient evidence to suggest that the true mean appraised values of the homes between the two appraisers are significantly different.

The Problem - Seeds in Soil Samples
Several methods of estimating the number of seeds in soil samples have been developed by ecologists. The paper "A Comparison of methods for Estimating Seed Numbers in the Soil" (*Journal of Ecology* (1990)1079-1093) considered three such methods. The accompanying data gives number of seeds detected by the direct method and by the standard method for 27 soil specimens.

Sample	1	2	3	4	5	6	7	8	9
Direct	24	32	0	60	20	64	40	8	12
Stratified	8	36	8	56	52	64	28	8	8

Sample	10	11	12	13	14	15	16	17	18
Direct	92	4	68	76	24	32	0	36	16
Stratified	100	0	56	68	52	28	0	36	12

Sample	19	20	2	22	23	24	25	26	27
Direct	92	4	40	24	0	8	12	16	40
Stratified	92	12	48	24	0	12	40	12	76

Does the data suggest that there are significant differences in the mean number of seeds detected by the two methods.
Follow these steps to perform the hypothesis test.

1. Enter data.

   Enter the observations for the Direct method into column C1. Name column C1 as Direct. Enter the observations for the Stratified method into column C2. Name column C2 as Stratified. Be sure to enter the observations as matched pairs.

2. Calculate the test statistic.

   Choose **Stat**>**Basic Statistics**>**Paired t**. . Place Direct in the First sample: text box. Place Stratified in the Second sample: text box. Choose Options. Accept the (default) level of 95.0 in the Confidence level: text box. Accept the (default) level of 0.0 in the Test mean: text box. Choose the (default) option of not equal in the Alternative: drop down dialog box. Choose **OK**.

The Minitab Output

Paired T-Test and CI: Direct, Stratified

Paired T for Direct - Stratified

	N	Mean	StDev	SE Mean
Direct	27	31.26	27.65	5.32
Stratified	27	34.67	28.87	5.56
Difference	27	-3.41	13.25	2.55

95% CI for mean difference: (-8.65, 1.84)
T-Test of mean difference = 0 (vs not = 0): T-Value = -1.34 P-Value = 0.193

Figure 10.5

Minitab again produces a number of descriptive statistics (N, Mean, StDev, SE Mean), as well as the observed t value of -1.34, and the two-tailed p-value of 0.193 (meaning p>α = .05). Since p is greater than α, the null hypothesis $H_0 : \mu_d \neq 0$ is not rejected. We thus do not have sufficient evidence to suggest that the true mean difference in the mean number of seeds detected by the two methods are significantly different.

Exercises

10.5 An experiment involving galvanic skin responses of subjects under two experimental treatments provided the following data.

Subject	1	2	3	4	5	6
Treatment1	3.9	3.7	3.6	3.9	4.1	4.5
Treatment2	3.1	3.6	2.7	3.4	3.1	3.2

Subject	7	8	9	10	11	12
Treatment1	2.9	3.6	4.1	3.4	3.5	3.5
Treatment2	2.1	2.6	3.2	3.1	3.2	2.6

Does the data suggest that there are significant differences in the galvanic skin responses between the two experimental treatments, at the .05 level of significance?

10.6 In a study of memory recall, eight people were given 10 min to memorize a list of 20 nonsense words. Each was asked to list as many of the words as he or she could remember both 1 hr and 24 hr later, as shown in the accompanying table.

Subject	1	2	3	4	5	6	7	8
1 hr later	14	12	18	7	11	9	16	15
24 hr later	10	4	14	6	9	6	12	12

Construct a 95% confidence interval on the true difference in mean number of words recalled after 1 hr and the number of words recalled after 24 hr.

10.7 In a study of the effectiveness of a drug in the sedation of wild animals, with the objective of safely relocating the animal to another location, blood samples were obtained at the time of capture and 30 minutes after injection. The levels of androgens at the time of capture and 30 minutes later are as follows:

Animal	1	2	3	4	5	6
At Capture	7.80	9.20	5.40	6.11	5.18	8.69
30 Minutes After Injection	15.11	12.41	6.51	19.85	6.12	9.21

Animal	7	8	9	10	11	12
At Capture	8.40	9.68	3.88	6.28	5.20	20.7
30 Minutes After Injection	5.95	9.00	6.64	4.20	5.04	88.73

Animal	13	14	15	16
At Capture	3.96	55.18	19.34	8.51
30 Minutes After Injection	8.22	81.73	44.00	9.05

Does the data suggest that there are significant differences in the levels of androgens in the blood between the time of capture and 30 minutes after the injection, at the .05 level of significance?

10.8 The effect of exercise on the amount of lactic acid in the blood was examined in the article "A Descriptive Analysis of Elite-Level Racquetball" (*Research Quarterly for Exercise and Sport* (1991):109-114). Eight men and seven women who were attending a week-long training camp participated in the experiment, and blood lactate levels were measured before and after playing three games of racquetball, as shown in the accompanying table.

Subject	M1	M2	M3	M4	M5	M6	M7	M8
Before	13	20	17	13	13	16	15	16
After	18	37	40	35	30	20	33	19

Subject	F1	F2	F3	F4	F5	F6	F7
Before	11	16	13	18	14	11	13
After	21	26	19	21	14	31	20

Does the data suggest that there are significant increases in the amount of lactic acid in the blood, at the .05 level of significance?

10.4 Two Population Proportions

New Minitab Commands

1. **Stat>Basic Statistics>2 Proportions** - Performs a test of two binomial proportions. Use the 2 Proportions command to compute a confidence interval and perform a hypothesis test of the difference between two proportions. For example, suppose you wanted to know whether the proportion of consumers who return a survey could be increased by providing an incentive such as a product sample. You might include the product sample with half of your mailings and see if you have more responses from the group that received the sample than from those who did not. In this section, you will use this command to perform a hypothesis test on the difference between two population proportions.

Hypothesis Tests on $\pi_1 - \pi_2$

Many investigations are carried out to compare the proportion of successes in one population (or resulting from one treatment) to the proportion of successes in a second populaiton (or from a second treatment). When comparing two populations or treatments on the basis of "success" proportions, it is common to focus on the quantity $\pi_1 - \pi_2$, the difference between the two proportions. Since p_1 provides an estimate of π_1 and p_2 provides an estimate of π_2, the obvious choice for an estimate of $\pi_1 - \pi_2$ is $p_1 - p_2$.

When the two random samples are selected independently of one another and both samples are large, we will use the test statistic:

$$z = \frac{(p_1 - p_2) - (\pi_1 - \pi_2)}{\sqrt{\frac{p_c(1-p_c)}{n_1} + \frac{p_c(1-p_c)}{n_2}}} \text{ , where } p_c = \frac{n_1 p_1 + n_2 p_2}{n_1 + n_2}.$$

A large sample confidence interval for $\pi_1 - \pi_2$ can easily be obtained by using

Minitab. The confidence interval for $\pi_1 - \pi_2$ is

$$\pi_1 - \pi_2 \pm (z \text{ critical value})\sqrt{\frac{p_1(1-p_1)}{n_1} + \frac{p_2(1-p_2)}{n_2}}$$

The Problem - Survey Responses

Many investigators have studied the effect of the wording of questions on survey responses. Consider the following two versions of a question concerning gun control.

1. Would you favor or oppose a law that would require a person to obtain a police permit before purchasing a gun? (favor = 1, oppose = 0)

2. Would you favor or oppose a law that would require a person to obtain a police permit before purchasing a gun, or do you think that such a law would interfere too much with the right of citizens to own guns? (favor = 1, oppose = 0)

   The extra phrase in question 2 that reminds individuals of the right to bear arms might tend to elict a smaller proportion of favorable responses that would the first question without the phrase. Does the data suggest that this is the case, at the .05 level of significance?

Follow these steps to perform the hypothesis test.

1. Open the worksheet.
   Choose **File**>**Open Worksheet**. Select the file a:survey.mtp. Choose **OK**.

2. Calculate the test statistic.
   Choose **Stat**>**Basic Statistics**>**2 Proportions**. . Darken the Samples in different columns: option button. Place Q1 in the First: text box. Place Q2 in the Second: text box. Choose Options. Accept the (default) level of 95.0 in the Confidence level: text box. Accept the (default) level of 0.0 in the Test mean: text box. Choose the greater than from the Alternative: drop down list box. Place a check in the Use pooled estimate of p for the test: checkbox. Choose **OK**. Choose **OK**.

The Minitab Output

Test and CI for Two Proportions: Q1, Q2

```
Success = 1

Variable        X       N   Sample p
Q1            463     615   0.752846
Q2            403     585   0.688889

Estimate for p(Q1) - p(Q2):  0.0639566
95% lower bound for p(Q1) - p(Q2):  0.0214152
Test for p(Q1) - p(Q2) = 0 (vs > 0):  Z = 2.47  P-Value = 0.007
```

Figure 10.6

Minitab produces a number of descriptive statistics (X, N, Sample p), as well as the observed z value of 2.47, and the one-tailed p-value of 0.007 (meaning p<α = .05). Since p is less than α, the null hypothesis $H_0 : \pi_1 - \pi_2 \neq 0$ is rejected. We thus do have sufficient evidence to suggest that the inclusion of the extra phrase about the right to bear arms does seem to result in fewer favorable responses than would be elicited without the phrase.

The Problem - Weed-killing Herbicides

Researchers at the National Cancer Institute released the results of a study that examined the effect of weed-killing herbicides on house pets. (Associated Press, September 4, 1991). The following data is conpatible with summary values given in the report. Dogs, some of whom were from homes where the herbicide was used on a regular basis were examined for the presence of malignant lymphoma.

Group	Sample Size	Number with lymphoma
Exposed	827	473
Unexposed	130	19

Does the data suggest that there are significant differences in the proportions of exposed dogs with lymphoma and unexposed dogs with lymphoma, at the 0.10 level of significance?.

Follow these steps to perform the hypothesis test and construct a 90% confidence interval on .

1. Calculate the test statistic.

   Choose **Stat**>**Basic** Statistics>**2 Proportions**. . Darken the Summarized data: option button. Place 827 in the First sample Trials: text box. Place 473 in the First sample Successes: text box. Place 130 in the Second sample Trials: text box. Place19 in the Second sample Successes: text box. Choose Options. Enter 90.0 in the Confidence level: text box. Accept the (default) level of 0.0 in the Test mean: text box. Choose not equal from the Alternative: drop down list box. Place a check in the Use pooled estimate of p for the test: checkbox. Choose **OK**. Choose **OK**.

The Minitab Output

Test and CI for Two Proportions

```
Sample     X      N  Sample p
1        473    827  0.571947
2         19    130  0.146154

Estimate for p(1) - p(2):  0.425793
90% CI for p(1) - p(2):  (0.367500, 0.484886)
Test for p(1) - p(2) = 0 (vs not = 0):  Z = 9.03  P-Value = 0.000
```

Figure 10.7

Minitab produces a number of descriptive statistics (X, N, Sample p), as well as the observed z value of 9.03, and the two-tailed p-value of 0.000 (meaning $p < \alpha = .05$). Since p is less than α, the null hypothesis $H_0 : \pi_1 - \pi_2 \neq 0$ is rejected. We thus do have sufficient evidence to suggest that the proportions of exposed dogs with lymphoma and unexposed dogs with lymphoma are significantly different.

   Exercises

10.9 A new insecticide is to be compared to an existing type of insecticide, where both types are designed to kill the same insect. Two areas of equal size are treated with the same amount of insecticide, one with the new insecticide, the other with the existing insecticide. Three hundred insects were initially released into each room, and after 1 hour, the numbers of dead insects are counted. The results appear below:

	New	Existing
Number of insects	300	300
Number of dead insectsw	285	135

Does the data suggest that the new insecticide is more effective than the existing insecticide, at the .05 level of significance?

10.10 Do teachers find their work rewarding and satisfying? The paper "Work-Related Attitudes" (Psychological Reports (1991):443-450) reported the results of a survey of 395 elementary school teachers and 266 high school teachers. Of the elementary school teachers, 224 said they were very satisfied with their jobs, whereas 126 of the high school teachers were very satisfied with their work. Based on this data, is it reasonable to conclude that the proporiton of very satisfied is different for elementary school teachers than it is for high school teachers. Test the appropriate hypotheses using a 0.05 significance level. Construct a 95% confidence interval on the true difference in mean number of words recalled after 1 hr and the number of words recalled after 24 hr.

10.5   The Mann-Whitney Rank Sum Test

New Minitab Commands

1. **Stat>Nonparametrics>Mann-Whitney** - Performs a two-sample rank test for the difference between two population medians, and calculates corresponding point estimate and confidence interval. (Sometimes called the two-sample Wilcoxon rank sum test.) In this section, you will use this command to calculate the rank sum statistic.

One approach to making inferences about $\mu_1 - \mu_2$ when n_1 and n_2 are small is to assume that the two population distributions are normal and then use the two-sample t test or confidence interval. In some situations, however, the normality assumption may not be reasonable.

Procedures that do not require any overly specific assumptions about the population distributions are said to be *distribution-free* or nonparametric. The Mann-Whitney Rank Sum test is one of these distribution-free techniques. The Mann-Whitney Rank Sum test is based on first ranking the observations from both samples together as one data set and then for each sample finding the sum of the ranks of the corresponding observations.

The Problem - Critical-Thinking

Students in a psychology course participated in a test for critical-thinking ability.

The scores for the female and male students appear below.

Females	87	91	40	102	67	73
	110	49	120	65	74	
Males	149	71	98	99	134	84
	34	76	68			

Assuming that the critical-thinking test score distributions for both females and males have the same shape and spread, use the Mann-Whitney Rank Sum test to test the hypothesis that there are no differences in the average critical thinking scores between females and males at the .05 level of significance.

Follow these steps to perform the hypothesis test.

1. Enter data.

   Enter the observations for Females into column C1. Name column C1 as Females. Enter the observations for Males into column C2. Name column C2 as Males.

2. Calculate the test statistic.

   Choose <u>S</u>tat><u>N</u>onparametrics><u>M</u>ann-Whitney. Place Females in the <u>F</u>irst Sample: text box. Place Males in the <u>S</u>econd Sample: text box. Choose the (default) option of not equal in the <u>A</u>lternative: drop down dialog box. Choose <u>O</u>K.

The Minitab Output

Mann-Whitney Confidence Interval and Test

Females	N=11	Median =		74.00
Males	N=9	Median =		84.00
Point estimate for ETA1 - ETA2 is				-9.00
95.2% Percent CI for ETA1 - ETA2 is			(-35.99, 19.99)	
W = 107.0				

Test of ETA1 - ETA2 vs ETA1 not = ETA2 is significant at 0.5433
Cannot reject at alpha = 0.05

Figure 10.6

Minitab uses the symbol W to denote the rank sum statistic and uses the terms ETA1 and ETA2 in place of μ_1 and μ_2. Minitab indicates that the null hypothesis is not rejected at the .05 level of significance. We thus do not have sufficient evidence to indicate that the true mean critical-thinking scores are different for females and males. Observe that Minitab also produces a 95.2% confidence interval on $\mu_1 - \mu_2$.

Exercises

10.11 A study focusing on particulate air pollution as a factor in premature mortality from heart and lung disease focused on the burn time for samples of oak and pine..

Oak	1.25	1.58	.69	.50	1.75	1.57	1.44	1.79
Pine	1.00	1.22	.75	1.42	1.54	1.35		

Assume that the distributions of burn times for both oak and pine have the same shape and spread. Use the Mann-Whitney Rank Sum test to test the hypothesis that there are no differences in the average burn times between oak and pine at the .05 level of significance.

10.12 The urinary fluoride concentration (ppm) was measured both for a sample of livestock that had been grazing in an area previously exposed to fluoride pollution and for a similar sample that had grazed in an unpolluted region.

Polluted	21.3	18.7	23.0	17.1	16.8	20.9	19.7
Unpolluted	14.2	18.3	17.2	18.4	20.0		

Assume that the distributions of urinary fluoride concentration for both grazing areas have the same shape and spread. Use the Mann-Whitney Rank Sum test to test the hypothesis that the true average fluoride concentration for livestock grazing in the polluted region is larger than for the unpolluted region at the .05 level of significance.

10.13 Emergency tire inflators were compared with respect to the length of time required to inflate a tire.

BrandA	22	20	28	23	17	16	20	18	32
BrandB	30	26	34	27	20	24	25	29	

Assume that the distributions of inflation times for both brands have the same shape and spread. Use the Mann-Whitney Rank Sum test to test the hypothesis that there are no differences between brands in the true average inflation times at the .05 level of significance

10.14 Two brands of graphing calculators were examined to determine the number of hours of use before the batteries failed.

CalculatorA	53	60	57	58	64	52	65	48	55
CalculatorB	47	54	40	50	47	51	45	44	42

Assume that the distributions of usage times for both brands have the same shape and spread. Use the Mann-Whitney Rank Sum test to test the hypothesis that there are no differences between brands in the true average inflation times at the .05 level of significance

Chapter 11
The Analysis of Categorical Data And Goodness of Fit Tests

11.1   Overview

Most of the techniques presented in earlier chapters are designed for numerical data. It is often the case, however, that information is collected on categorical variables such as political affiliation, sex, or college major. As with numerical data, categorical data sets can be univariate (consisting of observations on a single categorical variable), bivariate (observations on two categorical variables), or even multivariate. Minitab doesn't easily do goodness-of-fit tests; hence that test is not addressed in this chapter. After reading this chapter you should be able to

Perform a Chi-Squared Test for

1.   Homogeneity and Independence in a Two-Way Table

11.2   Tests for Homogeneity and Independence

New Minitab Commands

1. **Stat>Tables>Chisquare Test** - Does a chi-square test of association (non-independence) for the table of counts given in the specified columns. In this section, you will use this command to perform a chi-square test to determine if the category proportions are the same for all of the populations.

2. **Stat>Tables>Cross Tabulation** - Prints one-way or multi-way contingency tables and tables of statistics for associated variables, and displays the output in an easy-to-read format in the Session window. In this section, you will use this command to produce a contingency table.

Data resulting from observations made on two different categorical variables can also be summarized using a tabular format. As an example, suppose that patients of a particular hospital are classified by residence (rural, suburban, urban) and smoking status (never smoked, smoked in the past, smokes $\leq$ 1 pack/day, smokes > 1 pack/day). A research wishes to determine whether there is any relationship between residence and smoking status. Let x denote the variable residence and y denote the variable smoking status. A random sample of 200 patients is selected, and each individual is asked for his or her x and y values. The data set is bivariate and might initially be displayed as follows:

Individual	x value	y value
1	urban	past smoked
2	rural	never smoked
3	suburban	smokes ≤ 1
...		
200	urban	smokes > 1

Bivariate categorical data of this sort can most easily be summarized by constructing a two-way frequency or contingency table. This contingency table consists of a row for each possible x category and a column for each possible y category.

Comparing Two or More Populations

When the value of a categorical variable is recorded for member of separate random samples obtained from each population under study, the central issue is whether the category proportions are the same for all of the populations. The test procedure uses a chi-squared statistic that compares the observed counts to those that would be expected if there were no differences between the populations.

The Problem - Attitudes Toward Advertising

Until recently, a number of professions were prohibited from advertising. In 1977, the U.S. Supreme Court ruled that prohibiting doctors and lawyers from advertising violated their right to free speech. The paper "Should Dentists Advertise?" (*J. of Ad. Research* (June 1982):33-38) compared the attitudes of consumers and dentists toward the advertising of dental services. Separate samples of 101 consumers and 124 dentists were asked to respond to the following statement: "I favor the use of advertising by dentists to attract new patients." Possible responses were: strongly agree, agree, neutral, disagree, and strongly disagree. The data presented in the paper appears in the following contingency table.

	Response				
Group	Strongly Agree	Agree	Neutral	Disagree	Strongly Disagree
Consumers	34	49	9	4	5
Dentists	9	18	23	28	46

The authors were interested in determining whether the two groups - consumers and dentists - differed in their attitudes toward advertising.

Follow these steps to test the hypothesis H_0 : the true category proportions are the same for all the populations (homogeneity of populations).

1. Enter data.

   Enter the data for Strongly Agree in column C1. Name column C1 StronglyAgree. Enter the data for Agree in column C2. Name column C2 as Agree. Enter the data for Neutral in column C3. Name column C3 as Neutral. Enter the datafor Disagree in column C4. Name column C4 as Disagree. Enter the data for Strongly Disagree in column C5. Name column C5 as StronglyDisagree.

2. Calculate χ^2.

Choose **Stat**>**Tables**>**Chisquare Test**. Place StronglyAgree, Agree, Neutral, Disagree, and StronglyDisagree in the **C**olumns containing the table: text box. Choose **OK**.

The Minitab Output

Chi-Square Test

Expected counts are printed below observed counts

	Strongly	Agree	Neutral	Disagree	Strongly	Total
1	34	49	9	4	5	101
	19.30	30.08	14.36	14.36	22.89	
2	9	18	23	28	46	124
	23.70	36.92	17.64	17.64	28.11	
Total	43	67	32	32	51	225

```
Chi-Sq = 11.192 + 11.908 +  2.003 +  7.478 + 13.985 +
          9.116 +  9.699 +  1.632 +  6.091 + 11.391 =  84.496
DF = 4, P-Value = 0.000
```

Figure 11.1

The Minitab output (Figure 11.1) indicates the observed values, the expected counts, the marginal totals, the calculated Chi-squared statistic ($\chi^2 = 84.496$), the degrees of freedom (df=4) and the p-value (p=0.000). The p-value (p=0.000) $\leq \alpha$, so H_0 is rejected. There is strong evidence to support the claim that the proportions in the response categories are not the same for dentists and consumers.

The Problem - Drinking Behavior

Data on drinking behavior for samples of male and female students is similar to data that appeared in the paper "Relationship of Health Behaviors to Alcohol and Cigarette Use by College Students" (*J. of College Student Development* (1992):163-170). The data is contained in the file a:behavior.mtp. Does there appear to be a gender difference with respect to drinking behavior? (Note: low=1-7 drinks/week, moderate = 8 - 24 drinks/week, high = 25 or more drinks/week.)

Follow these steps to test the null hypothesis H_0 : True proportions for the four drinking levels are the same for males and females at the $\alpha = .01$ level of significance.

1. Retrieve the file.

Choose **File**>**Open Worksheet**. Select the file a:behavior.mtp. Choose **OK**.

2. Obtain the contingency table.

Choose **Stat**>**Tables**>**Cross Tabulation**. Place DrinkingStatus in the Classification variables: text box. Place Gender in the Classification variables: text box. Place a check in the Display Counts checkbox. Choose **OK**.

The Minitab Output

Tabulated Statistics

```
Rows: Drinking      Columns: Gender

              Male    Female      All

Moderate       300       173      473
High            63        16       79
Low            478       661     1139
None           140       186      326
All            981      1036     2017
```

Figure 11.2

The Minitab output (Figure 11.2) indicates the contingency table from the data file.
1. Calculate χ^2.

Choose Stat>Tables>Cross Tabulation. Place DrinkingStatus in the Classification variables: text box. Place Gender in the Classification variables: text
box. Remove the check in the Display Counts checkbox. Place a check in the
Chisquare analysis checkbox. Darken the Above and expected count option
button. Choose OK.

The Minitab Output

Tabulated Statistics

```
Rows: Drinking      Columns: Gender

              Male    Female      All

Moderate       300       173      473
            230.05    242.95   473.00

High            63        16       79
             38.42     40.58    79.00

Low            478       661     1139
            553.97    585.03  1139.00

None           140       186      326
            158.56    167.44   326.00

All            981      1036     2017
            981.00   1036.00  2017.00

Chi-Square = 96.526, DF = 3, P-Value = 0.000


    Cell Contents --
                   Count
                   Exp Freq
```

Figure 11.3

The Minitab output (Figure 11.3) indicates the observed values, the expected counts,
the marginal totals, the calculated Chi-squared statistic ($\chi^2 = 96.526$), the degrees
of freedom (df=3) and the p-value (p=0.000). The p-value (p=0.000) $\leq \alpha$, so H_0
is rejected. There is strong evidence to reject the claim that the proportions for the
four drinking levels are the same for males and females at the $\alpha = .01$ level of significance. Looking at the data, it appears that males tend to have higher drinking
levels than do females.

The Problem - Risk Taking Behaviors

The paper "Factors Associated with Sexual Risk-Taking Behaviors Among Adolescents (*J. Marriage and Family* (1994):622-632) examined the relationship between gender and contraceptive use by sexually active teens. Each person in a sample of sexually active teens was classified according to gender and contraceptive use (with three categories: rarely or never use, use sometimes or most of the time, and always use), resulting in a 3×2 contingency table. Data consistent with percentages given in the paper appears in the table.

	Gender	
Contraceptive use	Female	Male
Rarely/never	210	350
Sometimes/most times	190	510
Always	400	930

Follow these steps to test the hypothesis H_0 : Gender and contraceptive use are independent at the $\alpha = .05$ level of significance.

1. Enter data.

   Enter the data for Male(s) in column C1. Name column C1 Male. Enter the data for Female(s) in column C2. Name column C2 as Female.

2. Calculate χ^2.

   Choose **Stat**>**Tables**>**Chisquare Test**. Place Male and Female in the Columns containing the table: text box. Choose **OK**.

The Minitab Output

Chi-Square Test

```
Expected counts are printed below observed counts

            Female      Male      Total
      1        210       350        560
            224.00    336.00

      2        190       320        510
            204.00    306.00

      3        400       530        930
            372.00    558.00

   Total       800      1200       2000

Chi-Sq =    0.875 +   0.583 +
            0.961 +   0.641 +
            2.108 +   1.405 =  6.572
DF = 2, P-Value = 0.037
```

Figure 11.4

The Minitab output (Figure 11.4) indicates the observed values, the expected counts, the marginal totals, the calculated Chi-squared statistic ($\chi^2 = 6.572$), the degrees of freedom (df=2) and the p-value (p=0.037). The p-value (p=0.037) $\leq \alpha$, so H_0 is rejected. There is strong evidence to indicate an association between gender and contraceptive use.

Exercises

11.1 The accompanying data represent a survey from a particular worksite recording job type and smoking status.

	Job Type		
Smoking Status	Assembly Line	Secretary	Executive
Nonsmokers	53	34	31
Light Smokers	24	31	37
Heavy Smokers	23	35	32

Test the hypothesis that there is no association between job type and smoking status, using a .05 level of significance.

11.2 The authors of the paper "A survey of Parent Attitudes and Practices Regarding Under Age Drinking" (J. Youth and Adolescence (1995):315-334) conducted a telephone survey of parents with preteen and teenage children. One of the questions asked was "How effective do you think you are in talking to your children about drinking?". Responses are summarized in the accompanying table.

	Age of Children	
Response	Preteen	Teen
Very effective	126	149
Somewhat effective	44	41
Not at all effective or don't know	51	26

Using a .05 level of significance, carry out a test to determine whether there is an association between age of children and parental response.

11.3 Jail inmates can be classified into one of the following four categories according to the type of crime committed: violent crime, crime against property, drug offenses, and public-order offenses. Suppose that random samples of 500 male inmates and 500 female inmates are selected, and each inmate is classified according to type of offense. The data in the accompanying table is based on summary values given in the article "Profile of Jail Inmates (*USA Today*, April 25, 1991).

	Sex	
Type of Crime	Male	Female
Violent	117	66
Property	150	160
Drug	109	168
Public-order	124	106

Determine whether male and female inmates differ with respect to type of offense, using a .05 level of significance.

11.4 The paper "Color Associations of Male and Female Fourth-Grade School Children" (J. of Psych. (1988):383-388) asked children to indicate what emotion they associated with the color red. The response and sex of the child were noted, and the data is summarized in the file a:color.mtp. Obtain the contingency table, then using a .05 level of significance, carry out a test to determine whether there is an association between response and gender.

Chapter 12
Simple Linear Regression And Correlation: Inferential Methods

12.1　Overview

Regression and correlation were introduced in Chapter 4 as techniques for describing and summarizing bivariate data consisting of (x, y) pairs. For example, we considered a regression of the dependent variable $y =$ unemployment expenditure as a percentage of gross domestic product for the 22 OECD countries on the independent variable $x =$ unemployment rate. The equation of the least squares line was $\widehat{y} = 0.104 + 0.169x$. When $x = 7$ is substituted into this equation, the number 1.286 results. This number can be interpreted either as a point estimate of the average expenditure for all 22 OECD countries that have an unemployment rate of 7.0 or as a point prediction for the unemployment rate of 7.0 in a single country that has an unemployment rate of 7.0. In this chapter, we address inferential methods for this type of data, including a confidence interval (interval estimate) for a mean y value, and a prediction interval for a single y value. After reading this chapter you should be able to

1. Obtain the Least Squares Regression Line
   Obtain a point estimate of σ_e^2
   Obtain the coefficient of determination
2. Perform a Hypothesis Test Concerning β, the Slope
3. Obtain a Prediction Interval for a Single y value
4. Obtain a Confidence Interval for a Mean y value
5. Check the Model Adequacy

12.2   The Simple Linear Regression Model

New Minitab Commands (and some Minitab commands used previously)

1. **Stat>Regression>Regression** - Performs simple, polynomial regression, and multiple regression using the least squares method. In this section, you will use this command to determine the least square equation between two variables.
   a. **Options** - Permits various options: weighted regression, fit the model with/without an intercept, calculate variance inflation factors and the Durbin-Watson statistic, and calculate and store prediction intervals for new observations. In this section, you will use this command to make predictions using the least squares regression line.

2. **Stat>Regression>Fitted Line Plot** - Fits a simple linear or polynomial (second or third order) regression model and plots a regression line through the actual data or the log10 of the data. The fitted line plot shows you how closely the actual data lie to the fitted regression line. In this section, you will obtain a fitted line plot to illustrate how the estimated relationship fits the data in a simple linear regression model.

A deterministic relationship is one in which the value of y is completely determined by the value of an independent variable x. Such a relationship can be described using traditional mathematical notation such as $y = f(x)$, where $f(x)$ is a specified function of x. For example, $y = 3 + 2x$ is a deterministic relationship. However, the variables of interest are often not deterministically related. For example, the value of $y =$ unemployment expenditure as a percentage of gross domestic product is certainly not determined solely by $x =$ unemployment rate.
A description of the relationship between two variables x and y that are not deterministically related can be given by specifying a probabilistic model. The general form of an additive probabilistic model is

$$y = f(x) + e, \text{where } e = \text{ random deviation.}$$

The simple linear regression model is a special case of the general probabilistic model in which the deterministic function $f(x)$ is linear.
The simple linear regression model assumes that there is a line with slope β and vertical or y intercept α, called the true or population regression line. When a value of the independent variable x is fixed and an observtion on the dependent variable y is made,

$$y = \alpha + \beta x + e.$$

The Problem-House Size and Selling Price

As individuals prepare to purchase a home, one the major issues is the selling price of the home. An examination of recent listings from a variety of relators revealed the following data.

Square Footage	840	2372	1900	2096	1358	2663	941
Selling Price ($1000's)	59.9	159.9	82.9	219.0	41.9	242.5	64.9

Square Footage	1232	1900	1080	1440	1120	3300	980
Selling Price ($1000's)	43.0	113.9	54.0	85.0	19.9	289.9	74.5

Square Footage	1307	1440	1228
Selling Price ($1000's)	94.9	75.0	92.5

Follow these steps to determine the least squares equation between selling price and the square footage of the home and to predict the selling price when the square footage is 2000.

1. Enter data.

   Enter the data for the square footage in column C1. Name column C1 as SquareFootage. Enter the data for the SellingPrice in column C2. Name column C2 as SellingPrice.

2. Create a scatterplot.

   Choose **Graph>Plot.** Place SellingPrice in the Y Graph variables: text box. Place SquareFootage in the X Graph variables: text box.

3. Add a title.

   Choose **Annotation>Title.** Place the title: Selling Price vs. Square Footage in the first line of the Title text box.. Choose **OK**. Choose **OK**.

The Minitab Output

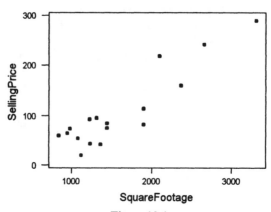

Figure 12.1

The scatter plot (Figure 12.1) strongly suggests the appropriateness of the simple linear regression model.

12.2 The Simple Linear Regression Model

1. Obtain the regression equation.
   Choose **Stat**>**Regression**>**Regression**. Place SellingPrice in the Response: text box. Place SquareFootage in the Predictors: text box.

2. Make a prediction.
   Choose Options. Place 2000 in the Prediction intervals for new observations: text box. Choose **OK**. Choose **OK**.

The Minitab Output

```
Regression Analysis

The regression equation is
SellingPrice = - 56.6 + 0.102 SquareFootage

Predictor      Coef        StDev          T        P
Constant     -56.60        20.93      -2.70    0.016
SquareFo     0.10206      0.01209       8.44    0.000

S = 32.91      R-Sq = 82.6%      R-Sq(adj) = 81.4%

Analysis of Variance

Source        DF          SS          MS        F        P
Regression     1       77175       77175    71.25    0.000
Error         15       16248        1083
Total         16       93423

    Fit  StDev Fit        95.0% CI              95.0% PI
 147.53       9.33   ( 127.63,  167.43)  (  74.59,  220.46)
```

Figure 12.2

The Minitab output (Figure 12.2) indicates the estimated regression line, SellingPrice = -56.6+0.102 SquareFootage ($\hat{y} = -56.6 + 0.102x$), as a well as a point estimate ($147.53) of the average selling price for all population members having a square footage of 2000 square feet.

Estimating σ^2 and σ

The value of σ determines the extent to which the observed points (x, y) tend to fall close to or far away from the population regression line. A point estimate of σ is based on $SS\,\mathrm{Re}\,sid = \sum(y - \hat{y})^2$, where $\hat{y}_1 = a + bx_1, ...\hat{y}_n = a + bx_n$ are the fitted or predicted y values and $y_1 - \hat{y}_1,...y_n - \hat{y}_n$ are the residuals. $SS\,\mathrm{Re}\,sid$ is a measure of the extent to which the sample data spreads out from the estimated regression line.

The statistic for estimating the variance σ^2 is

$$s_e^2 = \frac{SS\,\mathrm{Re}\,sid}{n - 2}$$

where

$$SS\,\mathrm{Re}\,sid = \sum(y - \hat{y})^2 = \sum y^2 - a\sum y - b\sum xy$$

The estimate of σ is the estimated standard deviation

$$s_e = \sqrt{s_e^2}$$

It is customary to call $n - 2$ the number of degrees of freedom associated with estmating σ^2 or σ in simple linear regression.

The coefficient of determination was defined earlier by

$$r^2 = 1 - \frac{SS\,\mathrm{Re}\,sid}{SSTo}$$

where

$$SSTo = \sum(y - \bar{y})^2 = \sum y^2 - \frac{(\sum y)^2}{n}$$

151

The value of r^2 can now be interpreted as the proportion of observed y variation that can be explained by (or attributed to) the model relationship.

The Problem - Shelf-life and Temperature

A particular medication comes in 8 fluid ounce bottles and the manufacturer recommends that after the bottle is unsealed it be kept under cool conditions. An examination of the shelf-life and the storage temperature revealed the following data.

Storage Temperature (x)	22	21	20	23	25
Shelf-life (y)	620	658	668	605	597

Storage Temperature (x)	24	14	26	17	15
Shelf-life (y)	619	780	584	703	750

Follow these steps to determine the least squares equation between shelf-life and the storage temperature of the medication and to predict the shelf-life when that storage temperature is 19

1. Enter data.

   Enter the data for the storage temperature in column C1. Name column C1 as Temperature. Enter the data for the shelf-life in column C2. Name column C2 as Shelf-life.

2. Create a scatterplot.

   Choose **Graph>Plot.** Place Shelf-life in the Y Graph variables: text box. Place Temperature in the X Graph variables: text box.

3. Add a title.

   Choose Annotation>Title. Place the title: Shelf-life vs. Temperatue in the first line of the Title text box.. Choose **OK.** Choose **OK.**

The Minitab Output

Shelf-life vs. Temperature

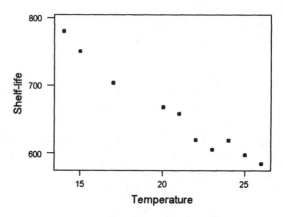

Figure 12.3

The scatter plot (Figure 12.3) strongly suggests the appropriateness of the simple linear regression model.

1. Obtain the regression equation.

   Choose <u>Stat</u>><u>Regression</u>><u>Regression</u>. Place Shelf-life in the
   <u>R</u>esponse: text box. Place Temperature in the Pre<u>d</u>ictors: text box.

2. Make a prediction.

   Choose Options. Place 19 in the Prediction <u>i</u>ntervals for new observations: text
   box. Choose <u>OK</u>. Choose <u>OK</u>.

The Minitab Output

Regression Analysis

```
The regression equation is
Shelf-life = 986 - 15.8 Temperature

Predictor       Coef        StDev           T        P
Constant      985.91        22.04       44.74    0.000
Temperat     -15.822        1.046      -15.13    0.000

S = 13.06      R-Sq = 96.6%      R-Sq(adj) = 96.2%

Analysis of Variance

Source         DF          SS          MS        F        P
Regression      1       39077       39077   228.95    0.000
Error           8        1365         171
Total           9       40442

    Fit   StDev Fit        95.0% CI              95.0% PI
 685.30        4.50   ( 674.92,  695.67)   ( 653.43,  717.17)
```

Figure 12.4

The Minitab output (Figure 12.4) indicates the estimated regression line, Shelf-life
= 986 - 15.8 Temperature ($\hat{y} = 986 - 15.8x$), as a well as a point estimate (685.3)
of the average shelf-life for all population members having a storage temperature
of 19. The Analysis of Variance table in the Minitab output contains SS for Error
= 1365, which is the $SS\,Re\,sid = 1365$. The SS for Total = 40442, which is noted
as $SSTo = 40442$. The estimated standard deviation, $s_e = \sqrt{s_e^2}$, can be found
by $\sqrt{MSError} = \sqrt{171} = 13.06$. Observe that $s_e = 13.06$ is also indicated on
the line with S, R-Sq and R-Sq(adj). The coefficient of determination, r^2, has a
value of 96.6%. Approximately 96.6% of the observed variation in shelf-life can
be attributed to the probabilistic linear relationship with storage temperature. The
magnitude of a typical sample deviation from the least squares line is about 13.06,
which is reasonably small in comparison to the y values themselves.

Exercises

12.1 The accompanying data on $x =$ treadmill run time to exhaustion (min) and $y =$
     20-km ski time (min) was taken from the paper "Physiological Characteristics
     and Performance of Top U.S. Biathletes" (Medicine and Science in Sports and
     Exercise (1995):1302-1310).

Treadmill Time(x)	7.7	8.4	8.7	9.0	9.6	9.6
Ski Time (y)	71.0	71.4	65.0	68.7	64.4	69.4

Treadmill Time (x)	10.0	10.2	10.4	11.0	11.7
Ski Time (y)	63.0	64.6	66.9	62.6	61.7

    a. Construct a scatterplot to determine if the simple linear regression model is appropriate. Add an appropriate title.

    b. Obtain the equation of the estimated regression line and predict the ski time for an individual whose treadmill time is 10 minutes.

    c. Construct a fitted line plot of the data.

    d. Refer to the Minitab output obtain in step (b) to estimate the average change in ski time associated with a one-minute increase in treadmill time.

    e. Refer to the Minitab output obtain in step (b) to obtain the value of r^2 and s_e. Write a brief interpretation of r^2 and s_e.

12.2 Using simple random sampling, a sample of size 16 was obtained with regards to x = number of hours studied and y = score (in percent).

Hours Studied (x)	2	12	12	14	5	14	3	2
Score (%) (y)	47	89	90	100	62	94	54	48

Hours Studied (x)	5	12	3	5	2	2	5	4
Score (%) (y)	62	89	59	58	52	48	59	67

    a. Construct a scatterplot to determine if the simple linear regression model is appropriate. Add an appropriate title.

    b. Obtain the equation of the estimated regression line and predict the score for an individual whose study time is 10 hours.

    c. Construct a fitted line plot of the data.

    d. Refer to the Minitab output obtain in step (b) to estimate the average change in score associated with a one-hour increase in study time.

    e. Refer to the Minitab output obtain in step (b) to obtain the value of r^2 and s_e. Write a brief interpretation of r^2 and s_e.

12.3 The accompanying data on x = age (in years) and y = systolic blood pressure were obtained for a random sample of 24 individuals.

Age (x)	35	52	35	55	68	40	33	48
Blood Pressure (y)	118	153	124	151	137	178	121	146

Age (x)	31	44	25	36	28	60	42	44
Blood Pressure (y)	113	137	102	128	114	163	131	138

Age (x)	63	49	59	55	42	69	45	55
Blood Pressure (y)	165	142	162	152	138	172	139	156

    a. Construct a scatterplot to determine if the simple linear regression model is appropriate. Add an appropriate title.

    b. Obtain the equation of the estimated regression line and predict the systolic blood pressure for an individual whose age is 40.

    c. Construct a fitted line plot of the data.

    d. Refer to the Minitab output obtain in step (b) to estimate the average change in systolic blood pressure associated with a one-year increase in age.

    e. Refer to the Minitab output obtain in step (b) to obtain the value of r^2 and s_e. Write a brief interpretation of r^2 and s_e.

12.4 The accompanying data on $x =$ advertising share and $y =$ market share for a particular brand of cigarettes during ten randomly selected years appeared in the paper "Testing Alternative Econometric Models on the Existence of Advertising Threshold Effect" (*J. of Marketing Research* (1984):298-308.

Advertising Share (x)	.103	.072	.071	.077	.086
Market Share (y)	.135	.125	.120	.086	.079

Advertising Share (x)	.047	.060	.050	.070	.052
Market Share (y)	.076	.065	.059	.051	.039

a. Construct a scatterplot to determine if the simple linear regression model is appropriate. Add an appropriate title.

b. Obtain the equation of the estimated regression line and predict the market share for an advertising share of .09.

c. Construct a fitted line plot of the data.

d. Refer to the Minitab output obtain in step (b) to estimate the average change in market share associated with a .01 increase in advertising share.

e. Refer to the Minitab output obtain in step (b) to obtain the value of r^2 and s_e. Write a brief interpretation of r^2 and s_e.

12.3   The Slope of the Population Regression Line

The slope β in the simple regression model is the average or expected change in the dependent variable y associated with a one-unit increase in the value of the independent variable x. For example, consider $x =$ storage temperature ($^\circ C$) and $y =$ shelf-life of the medication. Assuming that the simple linear regression model is appropriate for the population of medications, β would be the average decrease in shelf-life associated with a $1^\circ C$ increase in temperature.

Since the value of β is almost always unknown, it will have to be estimated from the sample data. The slope of the least squares line gives us a point estimate of β. As with any point estimate, though, it is desirable to have some indication of how accurately b estimates β. To proceed further, we need to know some information with regards to the sampling distribution of b.

Proberties of the Sampling Distribution of b

1.   The mean value of b is β. That is, $\mu_b = \beta$.

2.   The standard deviation of the statistic b is $\sigma_b = \frac{\sigma}{\sqrt{S_{xx}}}$

3.   The statistic b has a normal distribution.

The estimate standard deviation of b is

$$s_b = \frac{s_e}{\sqrt{S_{xx}}}$$

The probability distribution of the standardized variable

$$t = \frac{b - \beta}{s_b}$$

is the t distribution with $(n-2)$ degrees of freedom.

Hypothesis Tests Concerning β

Hypotheses about β can be tested using a t test similar to the t tests discussed in earlier chapters. The null hypotheseis states that β has a specific hypothesized value (usually 0). The test statistic results from standardizing b, the point estimate of β, under the assumption that the H_0 is true. When H_0 is true, the sampling distribution of this statistic is the t distribution with $(n-2)$ df. The test of $H_0 : \beta = 0$ versus $H_a : \beta \neq 0$ is often called the model utility test in the simple linear regression.

The Problem - pH and Formaldehyde

Durable-press cotton fabric is produced by a chemical reaction involving formaldehyde. For economic reasons, finished fabric usually receives it first wash at home rather than at the manufacturing plant. Because the pH of in-home wash water varies greatly from location to location, textile researchers are interested in how pH affects diferent fabric properties. The paper "Influence of pH in Washing on the Formaldehyde-Release Properties of Durable-Press Cotton (*Textile Research J.* (1981):263-270) reported the accompanying data (read from a scatter plot), where $x =$ washwater pH and $y =$ formaldehyde release (in ppm). Assume that a scatterplot has provided strong support for the simple linear regression model.

x	5.3	6.8	7.1	7.1	7.2	7.6	7.6	7.7	7.7
y	545	770	780	790	680	760	790	795	935

x	7.8	7.9	8.1	8.6	9.1	9.2	9.4	9.4	9.5
y	780	935	830	1015	1190	1030	1040	1250	1075

Follow these steps to determine whether a simple linear regression model would provide useful information for predicting the formaldehyde release from the washwater pH.

1. Enter data.
   Enter the data for the washwater pH (x) in column C1. Name column C1 as pH. Enter the data for the formaldehyde release in column C2. Name column C2 as Formaldehyde.

2. Obtain the regression equation.
   Choose **Stat>Regression>Regression.** Place Formaldehyde in the Response: text box. Place pH in the Predictors: text box.

3. Make a prediction.
   Choose Options. Place 7.5 in the Prediction intervals for new observations: text box. Choose **OK.** Choose **OK.**

The Minitab Output

Regression Analysis|

```
The regression equation is
Formaldehyde = - 311 + 151 pH

Predictor      Coef       StDev           T        P
Constant      -311.0      137.6        -2.26    0.038
pH             150.89      17.15         8.80    0.000

S = 78.01     R-Sq = 82.9%    R-Sq(adj) = 81.8%

Analysis of Variance

Source       DF        SS          MS        F        P
Regression    1     470979      470979    77.40    0.000
Error        16      97361        6085
Total        17     568340

    Fit   StDev Fit      95.0% CI           95.0% PI
  820.7        19.9    (  778.4,   863.0)  (  650.0,   991.4)
```

Figure 12.5

The Minitab output (Figure 12.5) indicates the estimated regression line, Formaldehyde = -311 + 151pH ($\hat{y} = -311 + 151x$), as a well as a point estimate (820.7) of the average formaldehyde release for all population members having a washwater pH of 7.5. The Minitab output (Figure 12.5) under the column labeled Coef includes the computed values of a ($a = -311.0$) and b ($b = 150.89$). The s_b appears under the column labeled StDev ($s_b = 17.15$). The value of the test statistic, the t ratio for the model utility test, is 8.80 and is found under the column labeled T. Since the associated p-value is less than .001, the H_0 is rejected. We conclude that there is a useful linear relationship between washwater pH and formaldehyde release.

The next line in the Minitab output includes s_e, ($s_e = 78.01$) and r^2 ($r^2 = 82.9\%$). The Analysis of Variance table includes the SSError or the $SS\,Resid$ ($SS\,Resid = 97361$), and the $SSTo$ ($SSTo = 568340$). Observe that $s_e = 13.06$ is also indicated on the line with S, R-Sq and R-Sq(adj). The coefficient of determination, r^2, has a value of 82.9%. Approximately 82.9% of the observed variation in shelf-life can be attributed to the probabilistic linear relationship with storage temperature. The magnitude of a typical sample deviation from the least squares line is about 78.01, which is reasonably small in comparison to the y values themselves.

Exercises

12.5 The paper "Effects of Enhanced UV-B Radiation on Ribulose-1,5-Biphosphate, Carboxylase in Pea and Soybean" (Environ. and Exper. Botany (1984):131-143) included the accompanying pea plant data, with y =sunburn index and x =distance (cm) from an ultrviolet light source.

Distance (x)	18	21	25	26	30	32	36	40
Sunburn Index (y)	4.0	3.7	3.0	2.9	2.6	2.5	2.2	2.0

Distance (x)	40	50	51	54	61	62	63
Sunburn Index (y)	2.1	1.5	1.5	1.5	1.3	1.2	1.1

Test the hypothesis $H_0 : \beta = 0$ versus $H_a : \beta \neq 0$ using a .05 level of significance. What does your conclusion say about the nature of the relationship between y =sunburn index and x =distance (cm) from an ultrviolet light source?

12.6 The accompanying table lists x = weight (in pounds) and y = height (in inches) of a sample of five-year-old boys.

Weight (x)	36	37	38	40	41
Height (y)	40	41	42	43	44

Weight (x)	47	48	49	43	46
Height (y)	48	40	52	45	46

Test the hypothesis $H_0 : \beta = 0$ versus $H_a : \beta \neq 0$ using a .05 level of significance. What does your conclusion say about the nature of the relationship between y =height (in inches) and x = weight (in pounds)?

12.7 The relationship between the depth of flooding and the amount of flood damage was examined in the paper "Significance of Location in Computing Flood Damage" (*J. of Water Resources Planning and Mgmt.* (1985):65-81. The accompying data on x = depth of flooding (feet abouve first-floor level) and y = flood damage (as a percent of structure value) was obtained using a sample of flood insurance claims.

Depth of Flooding (x)	1	2	3	4	5	6	7
Flood Damage (y)	10	14	26	28	29	41	43

Depth of Flooding (x)	8	9	10	11	12	13
Flood Damage (y)	44	45	46	47	48	49

Does the data suggest the existence of a positive linear relationship (one in which an increase in y tends to be associated with an increase in x) at the .05 level of significance?

12.8 The paper "Biomechanical Characteristics of the Final Approach Step, Hurdle, and Take-Off of Elite American Springboard Divers" (J. Human Movement Studies (1984):189-212) gave the accompanying data on y = judge's score and x = length of final step (m) for a sample of seven divers performing a forward pike with a single somersault.

Length of Final Step (x)	1.17	1.17	0.93	0.89	0.68	0.74	0.95
Judge's Score (y)	7.40	9.10	7.20	7.00	7.30	7.30	7.90

Test the hypothesis $H_0 : \beta = 0$ versus $H_a : \beta \neq 0$ using a .05 level of significance. What does your conclusion say about the nature of the relationship between y =judge's score and x = length of final step (m)?

12.4 Checking Model Adequacy

New Minitab Commands (and some Minitab commands used previously)

1. **Stat>Regression>Regression** - Performs simple, polynomial regression, and multiple regression using the least squares method. In this section, you will use this command to determine the least square equation between two variables.

a. **Graphs** - Displays residual plots. You do not have to store the residuals in order to produce these plots. In this section, you will use this command construct a standardized residual plot to determine the adequacy of the regression model.

The simple linear regression model equation is

$$y = \alpha + \beta x + e$$

where represents the random deviation of an observed y value from the population regression line $\alpha + \beta x$. Key assumption for the inferential methods presented in previous sections are

1. e has a normal distribution
2.   The standard deviation of e is σ, which does not depend on x.

Easily applied methods for checking the validity of the assumptions for the simple linear regression model are very desirable. When all the model assumptions are satisfied, the mean value of any residual is zero. Any observation that produces a very large positive or negative residual needs to be examined carefully for any anomalous circumstances, such as a recording error or exceptional experimental conditions. Identifying residuals with unusually large magnitudes is made easier by inspecting standardized residuals.

The Problem - Landslides!

Landslides are common events in tree-growing regions of the Pacific Northwest, so their effect on timber growth is of special concern to foresters. The paper "Effects of Landslide Erosion on Subsquent Douglas Fir Growth and Stocking Levels in the Western Cascades, Oregon" (*Soil Science Soc. of Amer. J.* (1984):667-671) reported on the results of a study in which growth in a landslide area was compared with growth in a previously clear-cut area. We present data on clear-cut growth, with x = tree age (years) and y = 5-year height growth (cm).

Tree Age (x)	5	9	9	10	10	11	11	12
Height Growth (y)	70	150	260	230	255	165	225	340

Tree Age (x)	13	13	14	14	15	15	18	18
Height Growth (y)	305	335	290	340	225	300	380	400

Follow these steps to:

(a) construct a scatter plot (fitted line plot) for this data,

(b) determine the estimated linear regression equation for this data, and

(c) construct a standardized residual plot to determine the adequacy of the simple linear regression model.

1. Enter data.
   Enter the data for the Tree Age (x) in column C1. Name column C1 as TreeAge. Enter the data for the Height Growth (y) in column C2. Name column C2 as HeightGrowth.

2. Obtain a fitted line plot.
   Choose **Stat**>**Regression**>**Fitted Line Plot**. Place HeightGrowth in the

159

Response (Y): text box. Place TreeAge in the Predictor (X): text box. Darken the Linear Type of Regression Model option button. Choose Options. Place 5-Year Height Growth vs. Tree Age in the Title: text box. Choose **OK.** Choose **OK.**

The Minitab Output

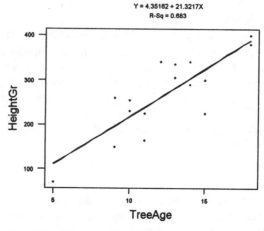

Figure 12.6

The fitted line plot, as shown in Figure 12.6, strongly supports the appropriateness of the simple linear regression model.

1. Obtain the regression equation.
    Choose **Stat**>**Regression**>**Regression.** Place HeightGrowth in the Response: text box. Place TreeAge in the Predictors: text box.

2. Obtain a standardized residual plot.
    Choose Graphs. Darken the Standardized Residuals for Plots: option button. Place a check in the Normal plot of residuals Residual Plots checkbox. Place a check in the Residuals vs fits Residual Plots checkbox. Place TreeAge in the Residuals versus the variables: text box. Choose **OK.** Choose **OK.**

The Minitab Output

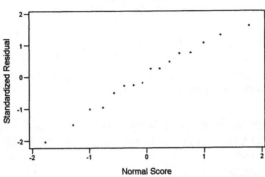

Figure 12.7

The Minitab output, as shown in Figure 12.7, indicates the normal probability plot of the standardized residuals. The plot casts no doubt on the normality assumption.

The Minitab Output

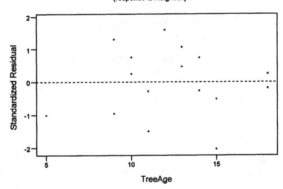

Figure 12.8

The Minitab output, as shown in Figure 12.8, indicates the standardized residual plot for tree age and 5-year growth. This plot show no unusual behavior that might call for model modifications or further analysis.

The Problem - Temperature and Fiberglass
The relationship between temperature and the workability of a fiberglass compound were explored, resulting in the following data.

Temperature (x)	42	40	45	41	43	46
Workability (y)	12.34	11.34	11.94	11.97	12.46	11.30

Temperature (x)	38	36	44	39	37
Workability (y)	9.33	6.30	12.33	10.46	7.94

Follow these steps to construct a standardized residual plot to determine the adequacy of the simple linear regression model.

1. Enter data.
   Enter the data for the Temperature (x) in column C1. Name column C1 as Temperature. Enter the data for the Workability (y) in column C2. Name column C2 as Workabilitity.

2. Obtain the regression equation.
   Choose **Stat**>**Regression**>**Regression**. Place Workability in the **R**esponse: text box. Place Temperature in the Pre**d**ictors: text box.

3. Obtain a standardized residual plot.
   Choose Graphs. Darken the **S**tandardized Residuals for Plots: option button. Place a check in the Residuals vs **f**its Residual Plots checkbox. Place Temperature in the Residuals versus the variables: text box. Choose **OK.** Choose **OK.**

The Minitab Output

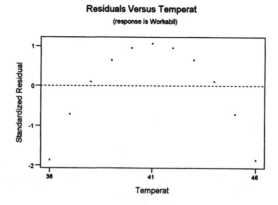

Figure 12.9

The Minitab output (Figure 12.9) indicates that the residual plot has a distinct curved pattern and therefore casts serious doubt on the suitability of the simple linear regression model. This plot shows behavior that might call for model modifications or further analysis.

Exercises

12.9 An investigation of the relationship between traffic flow x (thousands of cars per 24 hr) and lead content y of bark on trees near the highway ($\mu g/g$ dry weight) yielded the accompanying data.

Traffic Flow (x)	8.3	8.3	12.1	12.1	17.0
Lead Content (y)	227	312	362	521	640

Traffic Flow (x)	17.0	17.0	24.3	24.3	24.3
Lead Content (y)	539	728	945	738	759

a. Obtain a fitted line plot for this data.

b. Obtain the equation of the estimated regression line for prdicting the lead content from the traffic flow.

c. Obtain a standardized residual plot. Does the resulting plot suggest that a simple linear regression model is an appropraite choice? Explain your reasoning.

12.10 The following data set represents the number of years following graduation from a particular college and the amount of money (in thousands of dollars) contributed to the alumni association of the college.

Years (x)	1	2	3	4	6
Contributions (y)	823	835	1224	1361	2580

Years (x)	8	10	12	14	16
Contributions (y)	2540	2870	3870	5060	5020

a. Obtain a fitted line plot for this data.

b. Obtain the equation of the estimated regression line for prdicting the contributions from the number of years following graduation.

c. Obtain a standardized residual plot. Does the resulting plot suggest that a simple linear regression model is an appropraite choice? Explain your reasoning.

12.11 The authors of the paper "Age, Spacing and Growth Rate of Tamarix as an Indication of lake Boundary Fluctuations at Sebkhet Kelbia, Tunsia" (*J. of Arid Environ.* (1982):43-51) used a simple linear regression model tp describe the relationship between y = vigor (average width in centimeters of the last two annual rings) and x = stem density (stems/m^2). The estimated model was based on the following data.

Stem Density (x)	4	5	6	9	14
Vigor (y)	.75	1.20	.55	.60	.65

Stem Density (x)	15	15	19	21	22
Vigor (y)	.55	0	.35	.45	.40

a. What assumptions are required for the simple linear regression model to be appropriate?

b. Obtain a normal probability plot of the standardized residuals. Does the assumption that the random deviation distribution appear to be reasonable?

163

    c. Obtain a standardized residual plot. Are there any unusually large residuals?

    d. Is there anything about the standardized residual plot that would cause you to question the use of the simple linear regression model to describe the relationship between stem density and vigor?

12.12 The relationship between the depth of flooding and the amount of flood damage was examined in the paper "Significance of Location in Computing Flood Damage" (*J. of Water Resources Planning and Mgmt.* (1985):65-81. The accompying data on x = depth of flooding (feet abouve first-floor level) and y = flood damage (as a percent of structure value) was obtained using a sample of flood insurance claims.

Depth of Flooding (x)	1	2	3	4	5	6	7
Flood Damage (y)	10	14	26	28	29	41	43

Depth of Flooding (x)	8	9	10	11	12	13
Flood Damage (y)	44	45	46	47	48	49

    a. What assumptions are required for the simple linear regression model to be appropriate?

    b. Obtain a normal probability plot of the standardized residuals. Does the assumption that the random deviation distribution appear to be reasonable?

    c. Obtain a standardized residual plot. Are there any unusually large residuals?

    d. Is there anything about the standardized residual plot that would cause you to question the use of the simple linear regression model to describe the relationship between depth of flooding and flood damage?

12.5   Inferences Based on the Estimated Regression Line

New Minitab Commands (and some Minitab commands used previously)

1. <u>Stat</u>><u>Regression</u>><u>Regression</u> - Performs simple, polynomial regression, and multiple regression using the least squares method. In this section, you will use this command to determine the least square equation between two variables.

    a. **Options** - Allows you to perform weighted regression, fit the model with/without an intercept, calculate variance inflation factors and the Durbin-Watson statistic, and calculate and store confidence and prediction intervals for new observations. In this section, you will use this command obtain confidence and prediction intervals based on the extimated regression line.

The number obtained by substituting a particular x value x^* into the equation of the estimated regression line has two different interpretations. It is a point estimate of the average y value when $x = x^*$, and it is also a point prediction of a single y value to be observed when $x = x^*$. Properties of the sampling distribution are used to obtain both a confidence interval for $\alpha + \beta x^*$ and a prediction interval formula for a particular y observation. The width of the corresponding interval conveys information about the precision of the estimate or prediction. When the basic assumptions of the simple linear regression model are met, a confidence interval for $\alpha + \beta x^*$, the average y value when x has a value x^*, is $a + bx^* \pm (t_{critical\ value}) \cdot s_{a+bx^*}$, where the $t_{critical\ value}$ is based on $df = n - 2$. When the basic assumptions of the simple linear regression model are met, the

prediction interval for y^*, a single y observation when x has a value x^*, is $a + bx^* \pm (t_{critical\ value}) \cdot \sqrt{s_e^2 + s_{a+bx^*}^2}$ where the $t_{critical\ value}$ is based on $df = n - 2$. The prediction interval and confidence interval are centered at exactly the same

place, $a + bx^*$. The addition of under the square root symbol makes the prediction interval wider - often substantially so - than the confidence interval.

The Problem - Jaw Width of Sharks
Physical characteristics of sharks are of interest to surfers and scuba divers, as well as marine researchers. The data appearing in the magazines *Skin Diver* and *Scuba News*, on x = length (in feet) and y = jaw width (in inches) for 44 sharks are found in the file sharks.mtp.

Follow these steps to:

(a) construct a scatter plot (fitted line plot) for this data,

(b) determine the estimated linear regression equation for this data, and

(c) construct 90% confidence and prediction intervals for 15-foot-long sharks.

1. Open the worksheet.
   Choose **File>Open Worksheet**. Select the file a:sharks.mtp. Choose **OK**.

2. Create a scatterplot.
   Choose **Graph>Plot**. Place JawWidth(y) in the Y Graph variables: text box. Place Length(x) in the X Graph variables: text box.

The Minitab Output

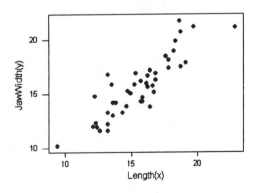

Figure 12.10

The scatter plot, as shown in Figure 12.10, strongly supports the appropriateness of the simple linear regression model.

3. Obtain the regression equation.

Choose Stat>Regression>Regression. Place JawWidth(y) in the Response: text box. Place Length(x) in the Predictors: text box. Choose **OK**.

The Minitab Output

```
The regression equation is
JawWidth(y) = 0.69 + 0.963 Length(x)

Predictor        Coef      SE Coef         T        P
Constant        0.688       1.299       0.53    0.599
Length(x      0.96345     0.08228      11.71    0.000

S = 1.376      R-Sq = 76.6%      R-Sq(adj) = 76.0%

Analysis of Variance

Source        DF         SS         MS         F        P
Regression     1     259.53     259.53    137.12    0.000
Residual Error  42      79.49       1.89
Total          43     339.02
```

Figure 12.11

The Minitab output (Figure 12.11) indicates the estimated regression line, JawWidth(y) = 0.69 + 0.963Length(x) ($\hat{y} = 0.69 + 0.963x$). The Minitab output under the column labeled Coef includes the computed values of a ($a = 0.688$) and b ($b = 0.96345$). The s_b appears under the column labeled StDev ($s_b = 0.08228$). The value of the test statistic, the t ratio for the model utility test, is 11.71 and is found under the column labeled T. Since the associated p-value is less than .001, the H_0 is rejected. We conclude that there is a useful linear relationship between JawWidth(y) and Length(x).

The next line in the Minitab output includes s_e, ($s_e = 1.376$) and r^2 ($r^2 = 76.6\%$). The Analysis of Variance table includes the SSError or the $SS\,Resid$ ($SS\,Resid = 79.49$), and the $SSTo$ ($SSTo = 339.02$). Observe that $s_e = 1.376$ is also indicated on the line with S, R-Sq and R-Sq(adj). The coefficient of determination, r^2, has a value of 76.6%. Approximately 76.6% of the observed variation in jaw widths can be attributed to the probabilistic linear relationship with length. The magnitude of a typical sample deviation from the least squares line is about 1.376, which is reasonably small in comparison to the y values themselves.

4. Obtain the confidence interval and prediction interval for 15-foot-long sharks.

Choose Stat>Regression>Regression. Place JawWidth(y) in the Response: text box. Place Length(x) in the Predictors: text box.

Choose Options. Place 15 in the Predicton intevals for new observations: text box.

Place 90 in the Confidence level: text box. Choose **OK**. Choose **OK**.

The Minitab Output

```
Predicted Values for New Observations

New Obs    Fit      SE Fit      90.0% CI           90.0% PI
1          15.140   0.213    ( 14.781, 15.498)  ( 12.798, 17.481)

Values of Predictors for New Observations

New Obs  Length(x)
 1         15.0
```

Figure 12.12

The Minitab output, as shown in Figure 12.12, contains a point estimate (Fit = 15.140) of the mean jaw width for 15-foot-long sharks, along with the standard error of the fitted value (SE Fit - 0.213).

The 90% confidence interval (90.0% CI) for the mean jaw width for sharks whose length is 15 ft. is between 14.781 and 15.498 inches. As with all confidence intervals, the 90% confidence level means that we have used a method to construct this interval estimate that has a 10% error rate.

The 90% prediction interval (90.0% PI) for the jaw width for an individual shark whose length is 15 ft. is between 12.798 and 17.481 inches. Observe that this 90% prediction interval is much wider than the 90% confidence interval when $x^* = 15$.

5. Obtain a fitted line plot.

Choose **Stat**>**Regression**>**Fitted Line Plot**. Place JawWidth(y) in the Response (Y): text box. Place Length(x) in the Predictor (X): text box. Darken the Linear Type of Regression Model option button. Choose Options. Place checks in both the Display confidence bands and Display prediction bands checkboxes. Place a title of Jaw Widths vs. Length of Sharks in the Title: text box. Choose **OK**. Choose **OK**.

The Minitab Output

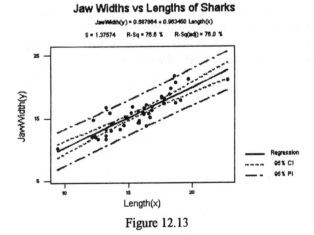

Figure 12.13

167

The fitted line plot, as shown in Figure 12.13, displays a scatterplot of the data, the estimated regression line and the confidence and prediction bands.strongly supports the appropriateness of the simple linear regression model.

Exercises

12.13 The paper"Effect of Temperature on the pH of Skim Milk" (J. of Dairy Research) (1988):277-280) reported on a study involving x = temperature (°C) under specified experimental conditions and y = milk pH. The accompanying data (read from a graph) is a representative subset of that which appeared in the paper.An investigation of the relationship between traffic flow x (thousands of cars per 24 hr) and lead content y of bark on trees near the highway ($\mu g/g$ dry weight) yielded the accompanying data.

Temperature (°C)(x)	4	4	24	24	25	38	38	40
Milk pH (y)	6.85	6.79	6.63	6.65	6.72	6.62	6.57	6.52

Temperature (°C)(x)	45	50	55	56	60	67	70	78
Milk pH (y)	6.50	6.48	6.42	6.41	6.38	6.34	6.32	6.34

a. Obtain a 95% confidence interval for $\alpha + \beta(40)$, the true average milk pH when the milk temperature is 40°C.

b. Obtain a 95% prediction interval for an individual for a single milk pH when the milk temperature is 40°C.

12.14 According to "Reproductive Biology of the Aquatic Salamander Amphiuma tridactylum in Louidiana" (*Journal of Herpetology*) 1999:100-105, the size of a female salamander's snout is correlated with the number of eggs in her clutch. The following data is consistent with summary quantities reported in the article.

Snout-Vent Length (x)	32	53	53	53	54	57	57
Clutch Size (y)	45	215	160	170	190	200	270

Snout-Vent Length (x)	58	58	59	63	63	64	67
Clutch Size (y)	175	245	215	170	240	245	280

a. Obtain a 95% confidence interval for $\alpha + \beta(40)$, the true clutch size for a salamander with a snout-vent length of 65.

b. Obtain a 95% prediction interval for an individual for clutch size for a salamander with a snout-vent length of 65.

Chapter 13
Multiple Regression Analysis

13.1 Overview

The general objective of regression analysis is to establish a useful relationship between a dependent variable y and one or more indpendent (predictor) variables. The simple linear regression model $y = \alpha + \beta x + e$ has been used successfully by many investigators in a wide variety of disciplines to relate y to a single predictor variable x. Most practical applications of regression analysis utilize models that are more complex than the simple linear regression model; in most problems more than one independent variable is needed in the regression model. For example, some variation in house prices may be attributed to the size of the house, but knowledge of house size alone would not enable one to accurately predict a home's value. Price is also determined to some extent by other variables, such as the number of bathrooms, the number of bedrooms and the age of the home.

In this chapter, we extend the regression methodology developed in previous chapters to multiple regression models. After reading this chapter you should be able to

1. Obtain the Least Squares Regression Equation for Two or More Predictor Variables
2. Fit a Model and Assess its Utility
3. Obtain a Prediction Interval for a Single y value
4. Obtain a Confidence Interval for a Mean y value
5. Obtain the "Best" Model Using a Stepwise Procedure

13.2   Multiple Regression Models

New Minitab Commands (and some Minitab commands used previously)

1. **Stat>Regression>Regression** - Performs simple, polynomial regression, and multiple regression using the least squares method. In this section, you will use this command to determine the least square equation between a dependent variable y to two or more predictor variables.

    a. **Options** - Permits various options: weighted regression, fit the model with/without an intercept, calculate variance inflation factors and the Durbin-Watson statistic, and calculate and store prediction intervals for new observations. In this section, you will use this command to make predictions using the least squares regression line where two or more predictor variables are present.

2. **Stat>Regression>Fitted Line Plot** - Fits a simple linear or polynomial (second or third order) regression model and plots a regression line through the actual data or the log10 of the data. The fitted line plot shows you how closely the actual data lie to the fitted regression line. In this section, you will obtain a fitted line plot to illustrate how the estimated relationship fits the data in a quadratic and cubic regression model.

3. **Stat>Anova>Interactions plot** - Draws a single interaction plot if 2 factors are entered. In this section, you will use this command to obtain an interaction plot where the variables that interact are included as a predictor variable in a multiple regression model.

A general additive multiple regression model, which relates a dependent variable y to k predictor variables $x_1, x_2, ...x_k$ is given by the model equation

$$y = \alpha + \beta_1 x_1 + \beta_2 x_2 + ... + \beta_k x_k + e$$

The random deviation e is assumed to be normally distributed with mean value 0 and variance σ^2 for any values of $x_1, x_2, ...x_k$. This implies that for fixed $x_1, x_2, ...x_k$ values, y has a normal distribution with variance σ^2 and

$$\left(\frac{\text{mean } y \text{ value for fixed}}{x_1, x_2, ...x_k \text{ values}}\right) = \alpha + \beta_1 x_1 + \beta_2 x_2 + ... + \beta_k x_k$$

The $\beta_i's$ are called population regression coefficients; each β_i can be interpreted as the true average change in y when the predictor x_i increases by one unit and the values of all the other predictors remain fixed. The deterministic portion $\alpha + \beta_1 x_1 + \beta_2 x_2 + ... + \beta_k x_k$ is called the population regression function.

The Problem - Predicting Income

Individuals in a certain state are surveyed about information with regards to yearly income ($1000's), educational level (on a scale of 1 to 20, 2 units equal 1 year of education beyond high school) and age. Consider the variables

$$y = \text{yearly income}$$
$$x_1 = \text{educational level}$$
$$x_2 = \text{age}$$

The data is as follows.

Education	Age	Income	Education	Age	Income
11	38	41	4	47	36
16	23	35	5	41	47
14	38	78	4	50	46
8	44	42	23	47	67
10	42	62	5	48	56
7	62	82	2	59	46
6	23	33	7	35	40
8	55	37	10	57	72
4	47	34	1	44	32
17	49	60	11	53	68
8	43	52	4	62	48
4	37	41	6	47	51
7	27	37	6	65	52
14	65	85	9	56	71
13	47	75	5	62	46

Follow these steps to obtain the least squares regression equation.

1. Enter data.
   Enter the data for education in column C1. Name column C1 as Education. Enter the data for age in column C2. Name column C2 as Age. Enter the data for income in column C3. Name column C3 as Income.

2. Obtain the regression equation.
   Choose Stat>Regression>Regression. Place Income in the Response: text box. Place Education in the Predictors: text box. Place Age in the Predictors: text box. (Education followed by Age)

3. Make a prediction.
   Choose Options. Place 10 (education) and 55 (age) in the Prediction intervals for new observations: text box. Choose OK. Choose OK.

Chapter 13

The Minitab Output

Regression Analysis

The regression equation is
Income = 0.98 + 1.89 Education + 0.758 Age

Predictor	Coef	StDev	T	P
Constant	0.980	9.610	0.10	0.920
Educatio	1.8934	0.4144	4.57	0.000
Age	0.7581	0.1773	4.28	0.000

S = 10.87   R-Sq = 56.4%   R-Sq(adj) = 53.2%

Analysis of Variance

Source	DF	SS	MS	F	P
Regression	2	4136.1	2068.0	17.50	0.000
Error	27	3191.1	118.2		
Total	29	7327.2			

Source	DF	Seq SS
Educatio	1	1975.1
Age	1	2160.9

Fit	StDev Fit	95.0% CI	95.0% PI
61.61	2.58	(56.32, 66.89)	(38.68, 84.54)

Figure 13.1

The Minitab output (Figure 13.1) indicates the regression equation is $y = 0.98 + 1.89x_1 + 0.758x_2$ (Income = 0.98+1.89 Education+0.758Age). For individuals whose education is 10 units and are 55 years old, the mean y value is 61.61 ($1000). The true average change in income when education increases by 1 unit and the other predictors remain fixed is $\beta_1 = 1.89$. Similarly, since $\beta_2 = 0.758$, when age increases by 1 year and the other predictors are fixed, we expect income to increase by 0.758 ($1000).

A Special Case: Polynomial Regression
The kth–degree polynomial regression model
$$y = \alpha + \beta_1 x + \beta_2 x^2 + \cdots + \beta_k x^k + e$$
is a special case of the general multiple regression model with $x_1 = x_1$, $x_2 = x^2$, ...$x_k = x^k$. The population regression function (mean value of y for fixed values of the predictors) is $\alpha + \beta_1 x + \beta_2 x^2 + \cdots + \beta_k x^k$. The mose important special case other than simple linear regression ($k = 1$) is the quadratic regression model
$$y = \alpha + \beta_1 x + \beta_2 x^2 + e$$
This model replaces the line of mean values $\alpha + \beta x$ in simple linear regression with a parabolic curve of mean values $\alpha + \beta_1 x + \beta_2 x^2$. If $\beta_2 > 0$, the curve opens upward, whereas if $\beta_2 < 0$, the curve opens downward.

The Problem - Tire Wear

Underinflated or overinflated tires increase tire wear and decrease gas mileage. One particular brand and model of a tire was examined at different pressures (pounds per square inch) with the following mileage results.

Pressure	Mileage (1000's)
29	27
30	31
31	35
32	37
33	38
34	35
35	33

Follow these steps to obtain the least squares quadratic regression equation.

1. Enter data.
   Enter the data for pressure in column C1. Name column C1 as Pressure. Enter the data for mileage in column C2. Name column C2 as Mileage.

2. Calculate the quadratic component.
   Choose **C**alc>**C**alculator. Place PresSquared in the **S**tore result in variable: text box. Place Pressure**2 in the **E**xpression: text box. Choose **OK**.

3. Obtain the regression equation.
   Choose **S**tat>**R**egression>**R**egression. Place Mileage in the **R**esponse: text box. Place Pressure and PresSquared in the Pre**d**ictors: text box. (Pressure followed by PresSquared)

4. Make a prediction.
   Choose Options. Place 32 (pressure) and 1024 ($32^2 = 1024$) in the Prediction intervals for new observations: text box. Choose **OK**. Choose **OK**.

The Minitab Output

Regression Analysis

```
The regression equation is
Mileage = - 813 + 52.1 Pressure - 0.798 PresSquared

Predictor       Coef      StDev          T        P
Constant     -813.00      77.87     -10.44    0.000
Pressure      52.083      4.880       10.67    0.000
PresSqua     -0.79762    0.07623     -10.46    0.000

S = 0.6986      R-Sq = 97.7%      R-Sq(adj) = 96.6%

Analysis of Variance

Source       DF         SS         MS        F        P
Regression    2     83.476     41.738    85.51    0.001
Error         4      1.952      0.488
Total         6     85.429

Source       DF     Seq SS
Pressure      1     30.036
PresSqua      1     53.440

   Fit   StDev Fit      95.0% CI              95.0% PI
36.905      0.403   ( 35.785,  38.025) ( 34.664,  39.145)
```

Figure 13.2

The Minitab output (Figure 13.2) indicates the regression equation is $y = -813 +$

Chapter 13

$52.1x - 0.798x^2$ (Mileage = -813+52.1 Pressure - 0.798 PresSquared). For a tire pressure of 32 pounds per square inch, the mean y value is 36.9 (1000) miles.

1. Obtain a fitted line plot.
   Choose **Stat**>**Regression**>**Fitted Line Plot**. Place Mileage in the Response (**Y**): text box. Place Pressure in the Predictor (**X**): text box. Darken the Quadratic Type of Regression Model option button. Choose **OK**.

The Minitab Output

Regression Plot

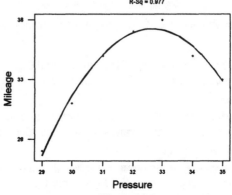

Figure 13.3

The Minitab output (Figure 13.3) indicates the scatterplot with the estimated regression line.

The Problem - Drug and Reaction Time

In a study involving the dosage of a drug and the reaction time of the individual to the drug, the following data were obtained.

Dosage (x)	6	6	8	8	10	10	12	12
Times (y)	7.6	6.9	4.1	4.5	4.0	3.2	1.5	1.8

Follow these steps to determine the least squares equation between Dosage (x) and Time (y).

1. Enter data.
   Enter the data for Dosage in column C1. Name column C1 as Dosage. Enter the data for the Time in column C2. Name column C2 as Time.
2. Obtain a scatterplot.
   Choose **Graph**>**Plot**. Place Time in the Y Graph variables: text box. Place Dosage in the X Graph variables: text box. Choose **A**nnotation>Title. Place the title: Dosage vs Reaction Times in the first line of the Title text box.. Choose **OK**. Choose **OK**. The scatterplot is shown in Figure 13.4.

The Minitab Output

Dosage vs Reaction Times

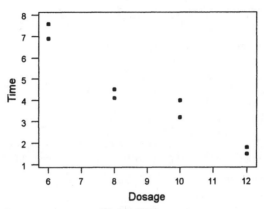

Figure 13.4

The scatterplot is shown in Figure 13.4.

3. Obtain the linear regression equation.
   Choose **Stat>Regression>Regression.** Place Time in the
   Response: text box. Place Dosage in the Predictors: text box. Choose **OK.**

The Minitab Output

Regression Analysis

The regression equation is
Time = 12.1 - 0.875 Dosage

Predictor	Coef	StDev	T	P
Constant	12.0750	0.9303	12.98	0.000
Dosage	-0.8750	0.1003	-8.72	0.000

S = 0.6344   R-Sq = 92.7%   R-Sq(adj) = 91.5%

Analysis of Variance

Source	DF	SS	MS	F	P
Regression	1	30.625	30.625	76.09	0.000
Error	6	2.415	0.403		
Total	7	33.040			

Figure 13.5

The Minitab output as shown in Figure 13.5 indicates the regression equation is
$y = 12.1 - 0.875x$ (Time = 12.1 - 0.875* Dosage) with $R^2 = 92.7\%$.

4. Introduce a squared term.
   Choose **Calc>Calculator.** Place DosageSqrd in the Store result in variable: text box. Place Dosage**2 in the Expression: text box. Choose **OK.**
5. Obtain the quadratic regression equation.
   Choose **Stat>Regression>Regression.** Place Time in the
   Response: text box. Place Dosage and DosageSqrd in the Predictors: text
   box. Choose **OK.**

The Minitab Output

Regression Analysis

```
The regression equation is
Time = 16.8 - 2.00 Dosage + 0.0625 DosageSqrd

Predictor       Coef        StDev          T         P
Constant      16.825        4.255       3.95     0.011
Dosage       -2.0000       0.9895      -2.02     0.099
DosageSq      0.06250      0.05470       1.14     0.305

S = 0.6189     R-Sq = 94.2%     R-Sq(adj) = 91.9%

Analysis of Variance

Source        DF          SS           MS          F         P
Regression     2      31.125       15.562      40.63     0.001
Error          5       1.915        0.383
Total          7      33.040
```

Figure 13.6

The Minitab output, as shown in Figure 13.6, indicates the regression equation is $y = 16.8 - 2.00x + 0.0625x^2$ (Time = 16.8 - 2.00 Dosage + 0.0625 DosageSqrd) with $R^2 = 94.2\%$.

6. Introduce a cubic term.

   Choose **Calc**>**Calculator**. Place DosageCubed in the Store result in variable: text box. Place Dosage**3 in the Expression: text box. Choose **OK**.

7. Obtain the cubic regression equation.

   Choose **Stat**>**Regression**>**Regression**. Place Time in the Response: text box. Place Dosage, DosageSqrd and DosageCubed in the Predictors: text box. Choose **OK**. The Minitab output, as shown in Figure 13.7, indicates the regression equation is $y = 64.6 - 19.1x + 2.03x^2 - 0.0729x^3$ (Time = 64.6 - 19.1 Dosage + 2.03 DosageSqrd - 0.0729 DosageCubed) with $R^2 = 97.9\%$.

The Minitab Output

Regression Analysis

```
The regression equation is
Time = 64.6 - 19.1 Dosage + 2.03 DosageSqrd - 0.0729 DosageCubed

Predictor       Coef        StDev          T         P
Constant       64.60        18.15       3.56     0.024
Dosage       -19.121        6.459      -2.96     0.042
DosageSq      2.0312       0.7397       2.75     0.052
DosageCu     -0.07292      0.02736      -2.66     0.056

S = 0.4153     R-Sq = 97.9%     R-Sq(adj) = 96.3%

Analysis of Variance

Source        DF          SS           MS          F         P
Regression     3      32.350       10.783      62.51     0.001
Error          4       0.690        0.173
Total          7      33.040

Source        DF       Seq SS
Dosage         1      30.625
DosageSq       1       0.500
DosageCu       1       1.225
```

Figure 13.7

8. Compare the Minitab output from each of the three regressions. Which regression equation "best fits" the data? How would you justify your response?

Interaction between Variables

The population regression function

$$\left(\begin{matrix}\text{mean } y \text{ value for fixed}\\ x_1, x_2, ...x_k \text{ values}\end{matrix}\right) = \alpha + \beta_1 x_1 + \beta_2 x_2 + ... + \beta_k x_k$$

exhibit a characteristic of all first-order models ($k = 1$). If you graph the mean value functions - say x_1- for fixed values of the other variables, the each graph will be a straight line. If you repeat the process for other values of the fixed independent variables, the lines will be parallel. This indicates that the effect on the mean y value of a change in x_1 is independent of the other variables in the model. When this situation occurs, we say that the independent variables in the model do not interact.

If a term involving the cross-product $x_1 x_2$ is added to the model, the effect on the mean y value is now dependent on the value of x_2. When this situation occurs, we say x_1 and x_2 interact. If you now graph the mean value functions -say x_1 - for fixed values of the other variables and repeat the process, the lines may no longer be parallel. When the slopes are different, the variables are said to interact.

To create an interaction variable using Minitab, use the **Calc>Calculator** command placing $x_1 x_2$ in the Store result in variable: text box. Place $x_1 * x_2$ in the Expression: text box. Choose **OK.**

If the change in the mean y values associated with a one-unit increase in one independent variable depends on the values of a second independent variable, there is interaction between these two variables. When the variables are denoted by x_1 and x_2, such interaction can be modeled by including $x_1 x_2$, the product of the variables that interact, as a predictor variable.

The general equation for a multiple regression model based on two independent variables x_1 and x_2 that also includes an interaction predictor is

$$y = \alpha + \beta_1 x_1 + \beta_2 x_2 + \beta_3 x_1 x_2 + e$$

The Problem - Burn Times

In an experiment designed to determine which of 2 different combustion systems is preferable, the burn time for 2 different types of fuels was measured.

	Fuel Type I	Fuel Type II
	37.4	33.0
System I	35.1	31.1
	41.8	27.1
	31.8	30.8
System II	28.5	36.9
	36.6	34.3

A simple numerical coding can incorporate qualitative (categorical) variables into this model. Let

$$x_1 = \begin{array}{l} 1 \text{ if the combustion system is System I} \\ 0 \text{ if the combustion system is System II} \end{array}$$

$$x_2 = \begin{array}{l} 1 \text{ if the fuel is Type I} \\ 0 \text{ if the fuel is Type II} \end{array}$$

The model is then

$$y = \alpha + \beta_1 x_1 + \beta_2 x_2 + \beta_3 x_1 x_2 + e$$

Follow these steps to obtain the least squares regression equation.

1. Enter data.

   Enter the data for the burn times in column C1. Name column C1 as BurnTime.
   Enter the coded data for the combustion systems in column C2. Name column
   C2 as X1. Enter the coded data for the fuel types in column C3. Name column
   C3 as X2. The data file should look like this.

BurnTime	X1	X2
374.	1	1
35.1	1	1
41.8	1	1
31.8	0	1
28.5	0	1
36.6	0	1
33.0	1	0
31.1	1	0
27.1	1	0
30.8	0	0
36.9	0	0
34.4	0	0

2. Set up the interaction term.

   Choose **Calc**>**Calculator**. Place X1X2 in the Store result in variable: text box.
   Place X1*X2 in the Expression: text box. Choose **OK**.

3. Obtain the regression equation.

   Choose **Stat**>**Regression**>**Regression**. Place BurnTime in the
   Response: text box. Place X1, X2 and X1X2 in the Predictors: text box.
   Choose **OK**.

The Minitab Output

Regression Analysis

```
The regression equation is
BurnTime = 34.0 - 3.63 X1 - 1.73 X2 + 9.43 X1X2

Predictor      Coef       StDev        T         P
Constant     34.033       1.972      17.26    0.000
X1           -3.633       2.788      -1.30    0.229
X2           -1.733       2.788      -0.62    0.551
X1X2          9.433       3.943       2.39    0.044

S = 3.415      R-Sq = 51.0%     R-Sq(adj) = 32.6%

Analysis of Variance

Source       DF        SS         MS       F        P
Regression    3       96.96      32.32    2.77    0.111
Error         8       93.31      11.66
Total        11      190.27

Source       DF      Seq SS
X1            1        3.52
X2            1       26.70
```

Figure 13.8

The Minitab output (Figure 13.8) indicates the regression equation is $y = 34.0 - 3.63x_1 - 1.73x_2 + 9.43x_1x_2$ (BurnTime = $34.0 - 3.63x_1 - 1.73x_2 + 9.43x_1x_2$).

1. Obtain an interaction plot.

   Choose <u>S</u>tat><u>A</u>nova><u>I</u>nteractions plot. Place X1 X2 in the <u>F</u>actors: text box. Place BurnTime in the Source of response data <u>R</u>aw response data in: text box. Choose <u>OK</u>.

The Minitab Output

Interaction Plot - Means for BurnTime

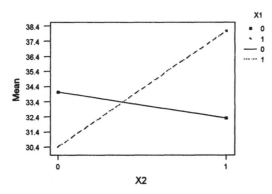

Figure 13.9

The Minitab output (Figure 13.9) indicates the interaction effects of x_1, the type of combustion system and x_2, the fuel type. The interaction plot indicates that the maximum burn time for fuel type II ($x_2 = 0$) is achieved by using combustion system II ($x_1 = 0$), while the maximum burn time for fuel type I ($x_2 = 1$) is achieved by using combustion system I ($x_1 = 1$). Observe that the slopes of the mean value functions are unequal.

Exercises

13.1 The cotton aphid poses a threat to cotton crops in Iraq. The accompanying data on

$$y = \text{infestation rate (aphids/100 leaves)}$$
$$x_1 = \text{mean temperature } (^0C)$$
$$x_2 = \text{mean relative humidity}$$

appeared in the paper "Estimation of the Economic Threshold of Infestation for Cotton Aphid" (Mesopotamia J. Ag. (1982):71-75). Find the estimated regression equation using the model

$$y = \alpha + \beta_1 x_1 + \beta_2 x_2 + e$$

y	x_1	x_2	y	x_1	x_2
61	21.0	57.0	77	24.8	48.0
87	28.3	41.5	93	26.0	56.0
98	27.5	58.0	100	27.1	31.0
104	26.8	36.5	118	29.0	41.0
102	28.3	40.0	74	34.0	25.0
63	30.5	34.0	43	28.3	13.0
27	30.8	37.0	19	31.0	19.0
14	33.6	20.0	23	31.8	17.0
30	31.3	21.0	25	33.5	18.5
67	33.0	24.5	40	34.5	16.0
6	34.3	6.0	21	34.3	26.0
18	33.0	21.0	23	26.5	26.0
42	32.0	28.0	56	27.3	24.5
60	27.8	39.0	59	25.8	29.0
82	25.0	41.0	89	18.5	53.5
77	26.0	51.0	102	19.0	48.0
108	18.0	70.0	97	16.3	79.5

13.2 The following data represent salaries (y) and years of experience (x) for a sample of employees in one particular job category from one field.

Years (x)	2	7	28	23	18	7
Salary (y)	23.4	23.2	80.3	66.7	50.7	27.0

Years (x)	15	13	15	25	4	28
Salary (y)	42.1	34.7	39.8	67.5	17.1	65.8

Follow these steps to determine the least squares equation between Years (x) and Salary (y).

  a. Enter data.

   Enter the data for Years in column C1. Name column C1 as Years. Enter the data for the Salary in column C2. Name column C2 as Salary.

  b. Obtain a scatterplot.

   Choose **Graph**>**Plot**. Place Salary in the Y Graph variables: text box. Place Years in the X Graph variables: text box. Choose Annotation>Title.

Place the title: Salary ($100) vs Years of Experience in the first line of the Title text box.. Choose **OK**. Choose **OK**.

c. Obtain the linear regression equation.
Choose **Stat**>**Regression**>**Regression**. Place Salary in the Response: text box. Place Years in the Predictors: text box. Choose **OK**.

d. Obtain a fitted line plot.
Choose **Stat**>**Regression**>**Fitted Line Plot**. Place Salary in the Response (Y): text box. Place Years in the Predictor (X): text box. Darken the Linear option button for the Type of Regression Model. Choose **OK**.

e. Introduce a squared term.
Choose **Calc**>**Calculator**. Place YearsSqrd in the Store result in variable: text box. Place Years**2 in the Expression: text box. Choose **OK**.

f. Obtain the quadratic regression equation.
Choose **Stat**>**Regression**>**Regression**. Place Salary in the Response: text box. Place Years and YearsSqrd in the Predictors: text box. Choose **OK**.

g. Obtain a fitted line plot for a quadratic equation.
Choose **Stat**>**Regression**>**Fitted Line Plot**. Place Salary in the Response (Y): text box. Place Years in the Predictor (X): text box. Darken the Quadratic option button for the Type of Regression Model. Choose **OK**.

h. Introduce a cubic term.
Choose **Calc**>**Calculator**. Place YearsCubed in the Store result in variable: text box. Place Years**3 in the Expression: text box. Choose **OK**.

i. Obtain the cubic regression equation.
Choose **Stat**>**Regression**>**Regression**. Place Salary in the Response: text box. Place Years, YearsSqrd and YearsCubed in the Predictors: text box. Choose **OK**.

j. Obtain a fitted line plot for a cubic equation.
Choose **Stat**>**Regression**>**Fitted Line Plot**. Place Salary in the Response (Y): text box. Place Years in the Predictor (X): text box. Darken the Cubic option button for the Type of Regression Model. Choose **OK**.

k. Which model, the linear, quadratic or cubic model appears to be most appropriate for describing the relationship between Salary and Years? Justify your answer.

13.3 Consider the following data obtained from an experiment examining the relationship between speed (x) and stopping distance.

Speed (x)	20	30	40	50	60	70	80
Stopping Distance (y)	45	75	120	200	275	380	500

Follow these steps to determine the model that "best fits" this data.

a. Enter data.
Enter the data for x in column C1. Name column C1 as Speed. Enter the data for y in column C2. Name column C2 as Distance.

b. Construct a scatterplot.
Choose **Graph**>**Plot**. Place Distance in the Y Graph variables: text box.

Place Speed in the X Graph variables: text box. Choose **OK**. What model does the scatterplot suggest? Is the model linear or quadratic?

c. Obtain a fitted line plot.
Choose **Stat**>**Regression**>**Fitted Line Plot**. Place Distance in the Response (Y): text box. Place Speed in the Predictor (X): text box. Darken the appropriate Type of Regression Model option button. Choose **OK**. Is the linear or quadratic model useful for describing the relationship between y and x?

13.4 A delivery service bases its charges for shipping a package on the distance the package is shipped and the weight of the package. The distance the package is shipped is coded into regions 1, 2, 3, etc.

Distance (x_1)	1	3	5	4	6
Weight (x_2)	6	3	4	7	1
Charge	2.75	4.00	8.50	9.70	4.90

Distance (x_1)	2	5	2	2	4
Weight (x_2)	1	7	5	1	8
Charge	2.00	15.00	2.40	1.50	14.50

Find the estimated regression equation using the model
$$y = \alpha + \beta_1 x_1 + \beta_2 x_2 + \beta_3 x_1 x_2 + e$$

13.5 An instructor keeps a record of attendance as well as a record of student's test scores. The final grades of 10 students, the number of absences and the first test scores are as follows:

Absences (x_1)	5	6	2	8	9
Score (x_2)	78	63	72	60	58
Final Grade (y)	90	88	93	79	77

Absences (x_1)	3	1	1	5	4
Score (x_2)	67	70	75	64	75
Final Grade (y)	88	97	98	84	94

Find the estimated regression equation using the model
$$y = \alpha + \beta_1 x_1 + \beta_2 x_2 + e$$

Predict the estimated final grade when the number of absences is 3 and the first test score is 72.

13.3   Fitting a Model and Assessing Its Utility

New Minitab Commands (and some Minitab commands used previously)

1. **Stat**>**Regression**>**Regression** - Performs simple, polynomial regression, and multiple regression using the least squares method. In this section, you will use this command to determine the least square equation between a dependent variable y to two or more predictor variables.

a. **Options** - Permits various options: weighted regression, fit the model with/without an intercept, calculate variance inflation factors and the Durbin-Watson statistic, and calculate and store prediction intervals for new observations. In this section, you will use this command to make predictions using the least squares regression line where two or more predictor variables are present.

Let's suppose a particular set of k predictor variables $x_1, x_2, ...x_k$ has been selected for inclusion in the model

$$y = \alpha + \beta_1 x_1 + \beta_2 x_2 + \cdots + \beta_k x_k + e$$

It is then necessary to estimate the model coefficients $\alpha, \beta_1, \beta_2, ...\beta_k$ and the regression function $\alpha + \beta_1 x_1 + \beta_2 x_2 + \cdots + \beta_k x_k$ (mean y value for specified values of the predictors), assess the model's utility, and perhaps use the estimated model to make further inferences.

The Problem - Soil and Sediment Adsorption

Soil and sediment adsorption, the extent to which chemicals collect in a condensed form on the surface, is an important characteristic because it influences the effectiveness of pesticides and various agricultural chemicals. The paper "Adsorption of Phosphates, Arsenate, Methanearsenate, and Cacodylate by Lake and Stream Sediments: Comparisons with Soils" (J. of Enviorn. Qual. (1984):499-504) presented the accompanying data consisting of $n = 13$ triples and proposed the model

$$y = \alpha + \beta_1 x_1 + \beta_2 x_2 + e$$

for relating

y =phosphate adsorption index
x_1 = amount of extractable iron
x_2 =amount of extractable aluminum

x_1 =iron	x_2 =aluminum	y =adsorption index
61	13	4
175	21	18
111	24	14
124	23	18
130	64	26
173	38	26
169	33	21
169	61	30
160	39	28
244	79	36
257	112	65
333	88	62
199	54	40

Follow these steps to obtain the least squares regression equation.

1. Enter data.

   Enter the data for extractable iron in column C1. Name column C1 as FE. Enter the data for extractable aluminum in column C2. Name column C2 as AL. Enter the data for the phosphate adsorption index column C3. Name column C3 as HPO. (Suggestion: save the data file for later use.)

2. Obtain the regression equation.

Choose **Stat>Regression>Regression.** Place HPO in the
Response: text box. Place FE in the Predictors: text box. Place AL in the
Predictors: text box. (FE followed by AL)

3. Make a prediction.
   Choose Options. Place 150 (FE) and 60 (AL) in the Prediction intervals for
   new observations: text box. Choose **OK.** Choose **OK.**

The Minitab Output

Regression Analysis

```
The regression equation is
HPO = - 7.35 + 0.113 FE + 0.349 AL

Predictor        Coef        StDev            T        P
Constant       -7.351        3.485        -2.11    0.061
FE            0.11273      0.02969         3.80    0.004
AL            0.34900      0.07131         4.89    0.000

S = 4.379       R-Sq = 94.8%      R-Sq(adj) = 93.8%

Analysis of Variance

Source          DF          SS           MS        F        P
Regression       2       3529.9       1765.0    92.03    0.000
Error           10        191.8         19.2
Total           12       3721.7

Source          DF       Seq SS
FE               1       3070.5
AL               1        459.4

  Fit  StDev Fit        95.0% CI              95.0% PI
30.50       1.92   (  26.21,   34.78)   (  19.84,   41.16)
```

Figure 13.10

The Minitab output (Figure 13.10) indicates the regression equation is $y = -7.35 + 0.113x_1 + 0.349x_2$ (HPO = -7.35 + 0.113 FE + 0.349 AL). The utility of the model can be assessed by examing the extent to which the predicted y values based on the estimated regression function are close to the y values actually observed. The p-value for the model utility test is found at the right hand side of the Analysis of Variance table (p=0.000 or p<0.001). In this case, the null hypothesis $H_0 : \beta_1 = \beta_2 = 0$ is rejected. The predicted mean y value for FE=150 and AL=60 is 30.5 which can be interpreted as either a point estimate for the mean value of HPO or as a point prediction for a single HPO value.

The coefficient of multiple determination, R^2, may be interpreted as the proportion of variation in observed y values that is explained by the fitted model. The coefficient of multiple determination, R^2, appears in the Minitab output as R-Sq = 94.8%. This would suggest that the chosen model has been very successful in relating y to the predictors.

Exercises

13.6 Refer to problem 13.1. Assess the utility of the model
$$y = \alpha + \beta_1 x_1 + \beta_2 x_2 + e$$

13.7 A study on the effect of applying fertilizer in bands is described in the paper "Fertilizer Placement Effects on Growth, Yield, and Chemical Composition of Burley Tobacco" (Agronomy J. (1984):183-188). The accompanying data with y = plant Mn (μg/g dry weight) and x = distance from the fertilizer band (cm) follows:

Distance (x)	0	10	20	30	40
Mn (y)	110	90	76	72	70

a. Find the estimated quadratic regression equation.

b. Perform the model utility test.

c. Interpret the values of R^2 and s_e.

13.8 A study involved the recording of the individual's age, nicotine intake and diastolic blood pressure.

Age (x_1)	72	31	72	64	53
Nicotine (x_2)	1.05	1.15	2.35	2.85	1.95
Diastolic b.p. (y)	90	63	117	111	106

Age (x_1)	22	49	66	37	45
Nicotine (x_2)	2.15	2.85	2.15	2.05	1.95
Diastolic b.p. (y)	75	113	106	92	108

a. Does the estimated equation specify a useful relationship between y and the independent variables. Assess the utility of the model
$$y = \alpha + \beta_1 x_1 + \beta_2 x_2 + e$$
using a .05 significance level.

b. Interpret the values of R^2 and s_e.

13.4   Inferences Based on an Estimated Model

New Minitab Commands (and some Minitab commands used previously)

1. **Stat>Regression>Regression** - Performs simple, polynomial regression, and multiple regression using the least squares method. In this section, you will use this command to determine the least square equation between a dependent variable y to two or more predictor variables.

    a. **Options** - Permits various options: weighted regression, fit the model with/without an intercept, calculate variance inflation factors and the Durbin-Watson statistic, and calculate and store prediction intervals for new observations. In this section, you will use this command to make predictions using the least squares regression line where two or more predictor variables are present.

In the previous section, we discussed estimating the coefficients $\alpha, \beta_1, \beta_2, ... \beta_k$ in the model $y = \alpha + \beta_1 x_1 + \beta_2 x_2 + \cdots + \beta_k x_k + e$ and then showed how the utility of the model could be confirmed by application of the F test for model utility. If the null hypothesis $H_0 : \beta_1 = \beta_2 = \cdots \beta_k = 0$ cannot be rejected at a reasonably small level of significance, it must be concluded that the model does not specify a useful relationship between y and any of the predictor variables $x_1, x_2, \ldots x_k$. The investigator must then search further for a model that does describe a useful relationship. Only if H_0 is rejected is it appropriate to proceed further with the chosen model and make inferences based on the estimated coefficients $a, b_1, \ldots b_k$ and the estimated regression function $a + b_1 x_1 + \ldots + b_k x_k$. This section addresses the issues of (1) drawing a conclusion about an individual regression coefficient β_i and (2) computing a point estimate or confidence interval for the corresponding mean y value or predicting a future y value with a point prediction or a prediction interval.

Inferences about Regression Coefficients
The Problem - Soil and Sediment Adsorption
Our analysis of the phosphate adsorption data introduced in the previous example has focused on the model $y = \alpha + \beta_1 x_1 + \beta_2 x_2 + e$ in which x_1 (extractable iron) and x_2 (extractable aluminum) affect the response seperately; that is, they do not interact. Suppose that the researcher wishes to investigate the possibility of interaction between x_1 and x_2. The proposed model is then

$$y = \alpha + \beta_1 x_1 + \beta_2 x_2 + \beta_3 x_1 x_2 + e$$

Let's explore that hypothesis that $H_0 : \beta_3 = 0$, that is, the interaction term does not belong in the model.

Follow these steps to test the hypothesis $H_0 : \beta_3 = 0$.

1. Enter data.

    Enter the data for extractable iron in column C1. Name column C1 as FE. Enter the data for extractable aluminum in column C2. Name column C2 as AL. Enter the data for the phosphate adsorption index column C3. Name column C3 as HPO.

2. Set up the interaction term.

    Choose **C̲alc>Calc̲ulator.** Place FEAL in the S̲tore result in variable: text box. Place FE*AL in the E̲xpression: text box. Choose **OK.**

3. Obtain the regression equation.

    Choose **S̲tat>R̲egression>Regression.** Place HPO in the R̲esponse: text box. Place FE in the Pre̲dictors: text box. Place AL in the Pre̲dictors: text box. Place FEAL in the Pre̲dictors: text box. (FE followed by AL followed by FEAL) Choose **OK.**

The Minitab Output

Regression Analysis

```
The regression equation is
HPO = - 2.37 + 0.0828 FE + 0.246 AL +0.000528 FEAL

Predictor       Coef        StDev          T        P
Constant      -2.368        7.179      -0.33    0.749
FE           0.08279      0.04818       1.72    0.120
AL            0.2460       0.1481       1.66    0.131
FEAL       0.0005278    0.0006610       0.80    0.445

S = 4.461      R-Sq = 95.2%      R-Sq(adj) = 93.6%

Analysis of Variance

Source        DF          SS           MS        F         P
Regression     3        3542.6       1180.9    59.34     0.000
Error          9         179.1         19.9
Total         12        3721.7

Source        DF      Seq SS
FE             1      3070.5
AL             1       459.4
FEAL           1        12.7
```

Figure 13.11

The Minitab output (Figure 13.11) indicates the value of $b_3 = 0.000528$ and the value of the test statistic comparing b_3 to 0, $t = \frac{b_3}{s_{b_3}} = 0.80$. The p-value associated with this test statistic is 0.445 and therefore the null hypothesis $H_0 : \beta_3 = 0$ is not rejected. The test statistic for β_3 -tests $H_0 : \beta_3 = 0$ when x_1 and x_2 are included in the model. The interaction predictor does not appear useful and should be removed from the model, as long as the other predictors are retained.

Inferences Based on the Estimated Regression Function

Figure 13.6 shows the Minitab output from fitting the model $y = \alpha + \beta_1 x_1 + \beta_2 x_2 + e$ to the phosphate data introduced earlier. The estimation and prediction information for FE=150 and AL=60 is contained at the bottom of Figure 12.6 and reproduced here.

```
   Fit  StDev Fit      95.0% CI            95.0% PI
 30.50       1.92   ( 26.21,  34.78)  ( 19.84,  41.16)
```

Figure 13.12

The estimated ($\hat{y}$) value for HPO is 30.5 with a estimated standard deviation ($s_{\hat{y}}$) of 1.92 as shown in Figure 13.12. The 95% confidence interval for the mean y value is 26.21 to 34.78, when FE=150 and AL=60. The 95% prediction for a single y value is 19.84 to 41.16, when FE=150 and AL=60.

Exercises

13.9 Refer to problem 13.4 concerning student's absences, first test score and final grade.

    a. Conduct tests for each of the following pairs of hypotheses.

        i. $H_0 : \beta_1 = 0$ versus $H_a : \beta_1 \neq 0$

        ii. $H_0 : \beta_2 = 0$ versus $H_a : \beta_2 \neq 0$

    b. Obtain a 95% confidence interval for the mean y value when $x_1 = 4$ and $x_2 = 73$. Interpret the resulting interval.

    c. Obtain a 95% prediction interval for a single y value when $x_1 = 4$ and $x_2 = 73$. Interpret the resulting interval.

13.10 Refer to problem 13.7 concerning an individual's age, incotine intake and diastolic blood pressure.

    a. Conduct tests for each of the following pairs of hypotheses.

        i. $H_0 : \beta_1 = 0$ versus $H_a : \beta_1 \neq 0$

        ii. $H_0 : \beta_2 = 0$ versus $H_a : \beta_2 \neq 0$

    b. Obtain a 95% confidence interval for the mean y value when $x_1 = 35$ and $x_2 = 2.00$. Interpret the resulting interval.

    c. Obtain a 95% prediction interval for a single y value when $x_1 = 35$ and $x_2 = 2.00$. Interpret the resulting interval.

13.5   Variable Selection

New Minitab Commands (and some Minitab commands used previously)

1. **Stat>Regression>Stepwise** - Identifies a useful subset of predictors for regression. Allows up to 100 predictors. The backwards elimination process begins with a model containing all possible predictors and removes them one at a time without reentering any. Ends when no variable in the model has an F-value less than F to remove. To do backwards elimination we will list all the predictors in the Enter text box, click the Options button, and set "F to enter" to 10000 (a value virtually impossible to obtain). In this section, you will apply the backwards elimination process where the initial model contains five predictor variables.

2. **Stat>Regression>Best Subsets** - Best subsets regression uses the maximum R^2 criterion. Suppose you specify m predictors, Minitab first selects the one-predictor regression model giving the largest R^2. Minitab then prints information on this model and the next best one-predictor model. Next Minitab finds the two-predictor model with the largest R^2, and prints information on it and the next best. The process continues until all m predictors are used. Best Subsets is an efficient way to select a group of "best subsets" for further analysis by selecting the smallest subset that fulfills certain statistical criteria. The subset model may actually estimate the regression coefficients and predict future responses with smaller variance than the full model using all predictors.

Suppose that an investigator has data on a number of predictor variables that might be incorporated into a model. The primary objective is then to select the set of predictors that in some sense specifies a "best" model.

Model selection methods can be divided into two types. There are those based on every possible model, computing one or more summary quantities from each fit, and comparing these quantities to identify the most satisfactory models. Minitab refers to this method as Best subsets regression. A second selection method is referred to as Stepwise regression. A backward stepwise procedure begins with all possible predictors in the model and deletes predictors one by one until all remaining predictors are judged important. A forward stepwise procedure begins with no predictors and then adds predictors until no predictor not in the model seems important. The "best" model contains relatively few predictors but has a large R^2 value and is such that no other model containing more predictors gives much of an improvement in R^2.

The Problem - Slash Pine Wood
An Example using Backward Stepwise Regression

The paper "Anatomical Factors Influencing Wood Specific Gravity of Slash Pines and the Implications for the Development of a High-Qualtity Pulpwood" (*TAPPI* (1964):401-444) reported the results of an experiment in which 20 specimens of slash pine wood were analyzed. A primary objective was to relate wood specific gravity (y) to various other wood characteristics. The independent variables on which observations were made were

$$x_1 = \text{number of fibers/mm}^2 \text{ in springwood}$$
$$x_2 = \text{number of fibers/mm}^2 \text{ in summerwood}$$
$$x_3 = \text{springwood \%}$$
$$x_4 = \text{\% springwood light absorption}$$
$$x_5 = \text{\% summerwood light absorption}$$

The data is stored in the file a:pulpwood.mtp.

Follow these steps to select the most satisfactory model.

1. Obtain the data file.
   Choose **File>Open Worksheet.** Select the file a:pulpwood.mtp. Observe that the columns are named X1, X2, X3, X4, X5 and Y.

2. Perform the stepwise regression procedure.
   Choose **Stat>Regression>Stepwise.** Place Y in the Response: text box. Place X1, X2, X3, X4 and X5 in the Predictors: text box. Place X1, X2, X3, X4 and X5 in the Enter: text box.

3. Apply the backward elimination procedure.
   Place 10000 in the F to enter: text box. Choose **OK.** Choose **OK.**

The Minitab Output

Stepwise Regression

F-to-Enter: 10000.00   F-to-Remove: 4.00

Response is   Y   on 5 predictors, with N =   20

Step	1	2	3	4
Constant	0.4421	0.4384	0.4381	0.5179
X1	0.00011	0.00011	0.00012	
T-Value	1.17	1.95	1.98	
X2	0.00001			
T-Value	0.12			
X3	-0.00531	-0.00526	-0.00498	-0.00438
T-Value	-5.70	-6.56	-5.96	-5.20
X4	-0.0018	-0.0019		
T-Value	-1.63	-1.76		
X5	0.0044	0.0044	0.0031	0.0027
T-Value	3.01	3.31	2.63	2.12
S	0.0180	0.0174	0.0185	0.0200
R-Sq	77.05	77.03	72.27	65.50

Figure 13.13

The Minitab output (Figure 13.13) indicates the results of the backward elimination procedure. The default value of F to remove the variable from the model is 4 (T-Value=2). The $t-value$ closest to zero when the model with all five predictors was fit was 0.12, so the corresponding predictor x_2 was removed from the model, as shown in the vertical column under Step 2. When the model with the four remaining predictors was fit, the $t-value$ closest to zero was -1.76, so the corresponding predictor x_4 was removed from the model. The next predictor to be dropped was x_1, since it's $t-value$ is 1.98. The final model contains only the predictors x_3 and x_5. This would suggest that the most satisfactory model is $y = 0.5179 - 0.00438x_3 + 0.0027x_5$. However, x_1 just barely met the elimation criterion, so the model with predictors x_1, x_3, and x_5 should be given serious consideration. You might want to compare the model containing x_1, x_3, and x_5 with the model containing just x_3, and x_5.

The Problem - Slash Pine Wood
An Example using Best subsets Regression

The paper "Applying Stepwise Multiple Regression Analysis to the Reaction of Formaldehyde with Cottong Cellulose" (*Textile Research J.* (1984):157-165) reported the results of an experiment involving wrinkle resistance and a number of independent variables. The dependent variable

$$y = \text{durable press ration}$$

is a quantitative measure of wrinkle resistance. The four independent variables,

190

and the abbreviations, used in the model building process are

$x_1 =$ HCHO (formaldehyde) concentation (con)

$x_2 =$ catalyst ratio (cat)

$x_3 =$ curing temperature (temp)

$x_4 =$ curing time (time)

In addition to the four independent variables, the investigators considered as potential predictors x_1^2 (consqd), x_2^2 (catsqd), x_3^2 (tempsqd), x_4^2 (timesqd), and all six interactions $x_1 x_2$ (con*cat), $x_1 x_3$ (con*temp), $x_1 x_4$ (con*time), $x_2 x_3$ (cat*temp), $x_2 x_4$ (cat*time), and $x_3 x_4$ (temp*time). The data is stored in the file a:frmldhyd.mtp. **Follow these steps** to select the most satisfactory model.

1. Obtain the data file.

   Choose **File**>**Open Worksheet.** Select the file a:frmldhyd.mtp. Choose **Open.** Observe that the columns are named con, cat, temp, time, consqd, catsqd, etc.

2. Perform the best subsets regression procedure.

   Choose **Stat**>**Regression**>**Best Subsets.** Place Y in the Response: text box. Place all predictors con-'temp*time' in the **F**ree predictors: text box. Choose **Options.** Place 4 in the M**i**nimum: Free predictors in each model text box. Choose **OK.** Choose **OK.**

The Minitab Output

```
Best Subsets Regression
Response is y

                                                c c c c t
                                            t t c o o a a e
                                        c c e i o n n t t m
                                        o a m m n * * * * p
                                    t t n t p e * t t t t *
                         R-Sq       c c e i s s s s c e i e i t
                                    o a m m q q q q a m i o i i
Vars  R-Sq  (adj)  C-p    S         n t p e d d d d t p e p e m

  4   82.2  79.4   13.9  0.63607    X   X   X         X
  4   82.1  79.2   14.1  0.63843    X       X   X     X
  5   86.7  83.9    7.4  0.56196    X   X X           X X
  5   85.5  82.5    9.6  0.58571    X   X X X         X
  6   88.5  85.5    5.9  0.53337    X     X X         X X X
  6   88.2  85.1    6.5  0.54094    X   X X X         X           X
  7   89.9  86.7    5.2  0.51045    X X X X X X       X
  7   89.9  86.7    5.3  0.51140    X X   X X X X     X
  8   91.5  88.3    4.2  0.47933    X X X X X X       X           X
  8   90.6  87.0    6.0  0.50535    X X   X X X X X   X
  9   91.7  87.9    5.9  0.48650    X X X X X X X     X           X
  9   91.6  87.8    6.0  0.48853    X X X X X X       X       X   X
 14   92.1  84.8   15.0  0.54639    X X X X X X X X X X X X X X
```

Figure 13.14

The Minitab output, as shown in Figure 13.14, indicates that the choice of a best model is not clear-cut. We certainly don't see the benefit of including more than $k = 8$ predictor variables (after that, adjusted R^2 begins to decrease) nor would we suggest a model with fewer than five predictor variables (adjusted R^2 is still increasing and CP is large). Based on this output, the best six-predictor model is a reasonable choice. That model is indicated by the largest adjusted R^2 within the six-predictor models (85.5). This model also has the smallest C-p statistic (5.9). A small value of Cp indicates that the model is relatively precise (has small variance) in estimating the true regression coefficients and predicting future responses. This precision will not improve much by adding more predictors. Models with considerable lack of fit have values of Cp larger than p (the number of parameters in the model).

Based upon this output, the best six-predictor model contains the variables, x_2 (cat), x_1^2 (consqd), x_2^2 (catsqd), x_1x_3 (con*temp), x_1x_4 (con*time), and x_2x_3 (cat*temp).

Follow these steps to determine the estimated regression equation.

1. Obtain the regression equation.

Choose <u>S</u>tat><u>R</u>egression><u>R</u>egression. Place y in the Response: text box. Place cat, consqd, catsqd, con*temp, con*time, and cat*temp in the Pre<u>d</u>ictors: text box. Choose <u>OK</u>.

The Minitab Output

Regression Analysis

```
The regression equation is
y = - 1.22 + 0.960 cat - 0.0373 consqd - 0.0389 catsqd
          + 0.00368 con*temp + 0.0193 con*time
          - 0.00128 cat*temp
```

Figure 13.15

The Minitab output, as shown in Figure 13.15, indicates the best six-predictor es-itmated regression equation.

Exercises

13.11 The $n = 25$ observations on $y =$ catch at intake (number of fish), $x_1 =$ water temperature (^0C), $x_2 =$ minimum tide height (m), $x_3 =$ number of pumps running, $x_4 =$ speed (knots), and $x_5 =$ wind-range direction (degrees) constitute a subset of the data that appeared in the paper "Multiple Regression Analysis for Forecasting Critical Fish Influxes at Power Station Intakes" (*J. Applied Ecol.* (1983):33-42). The data is contained in the file a:fish.mtp. Apply the backward elimination procedure to formulate a model for this data set.

13.12 The $n = 22$ observations on $y =$ diastolic blood pressure, $x_1 =$ age of individual, $x_2 =$ height of individual, $x_3 =$ weight of individual and $x_4 =$ nicotine intake are contained in the file a:diastolc.mtp. Apply the backward elimination procedure to formulate a model for this data set.

13.13 A subset of the data that appeared in the paper "Multiple Regression Analysis for Forecasting Critical Fish Influxes at Power Station Intakes" (*J. Applied Ecol.* (1983):33-42) is contained in the file a:catch.mtp. The dependent variable is

$y =$ catch at intake (number of fish).

The predictor variables are

$x_1 =$ water temperature (^oC)
$x_2 =$ minimum tide height (m)
$x_3 =$ number of pumps running
$x_4 =$ speed (knots)
$x_5 =$ wind-range of direction (degrees)

Use the Best subsets procedure to formulating a model.

Chapter 14
The Analysis of Variance

14.1 Overview

Methods for testing $H_0 : \mu_1 - \mu_2 = 0$, where μ_1 and μ_2 are the means of two different populations were discussed in Chapter 10. Many investigations involve a comparison of more than two population or treatment means. For example, an investigation carried out to study the nutritional value of cereal might report data on a number of different attributes that might affect the nutritional value, including type of cereal (wheat, bran, corn, etc.) Here is an example of the type of summary information that might be reported for calories.

Type of Cereal	Wheat	Bran	Corn	Sugar
Sample Size	15	15	12	22
Sample Mean	162.44	173.78	131.99	136.33

Let μ_1, μ_2, μ_3, and μ_4 denote the true average (i.e., population mean) calories for the types of cereal (wheat, bran, corn, and sugar), respectively. Does the data support the claim that $\mu_1 = \mu_2 = \mu_3 = \mu_4$, or does it appear that at least two of the μ's are different from one another. This is an example of a single-factor analysis of variance (ANOVA) problem, in which the objective is to decide whether the means for more than two populations or treatments are identical. In this chapter, we address situations involving analysis of variance (ANOVA). After reading this chapter you should be able to

1. Obtain the Single-Factor Analysis of Variance (ANOVA) Table for
   Unstacked Data and
   for Stacked Data
2. Perform the Tukey-Kramer Multiple Comparison Procedure
3. Perform ANOVA for a Randomized Block Experiment
4. Perform a Two-Factor ANOVA with Interaction

14.2 Single-Factor ANOVA

New Minitab Commands

1. **Stat>ANOVA>Oneway (Unstacked)** - Performs a one-way analysis of variance, with each group in a separate column. Data contained in seperate columns is referred to by Minitab as unstacked data. In this section, you will use this command to perform a one-way analysis of variance on data contained in four columns.

    a. **Graphs** - Displays boxplots and dotplots for each sample contained in a sample column. In this section, you will use this option, to construct comparative boxplots for the four samples (columns) of data.

2. **Stat>ANOVA>Oneway** - Performs a one-way analysis of variance, with the dependent variable in one column, subscripts in another. Data corresponding to different treatments contained in one column, with subscripts identifing the treatments in a seperate column, is referred to by Minitab as stacked data. In this section, you will perform a one-way analysis of variance on four samples, where all of the data is contained in one column, and the subscripts in another column.

Analysis of variance is used to compare data from several populations. When two or more populations or treatments are compared, the characteristic that distinguishes the population or treatments from one another is called the factor under investigation. For example, an experiment might be carried out to compare three different methods for teaching reading (three different treatments), in which case the factor of interest would be teaching method. This is a qualitative factor. If growth of fish raised in waters having different salinity levels - 0%, 10%, 20% and 30% - is of interest, the factor salinity level is quantitative.

A single-factor analysis of variance, ANOVA, problem involves a comparison of k population or treatment means $\mu_1, \mu_2, ...\mu_k$. The objective is to test

$$H_0 : \mu_1 = \mu_2 = ... = \mu_k$$

against

$$H_a : \text{at least two of the } \mu\text{'s are different.}$$

The analysis is based on k independently selected random samples, one from each population or for each treatment. A comparison of treatments based on independently selected experimental units is often referred to as a completely randomized design.

The Problem - Compression Strength

The paper "Compression of Single-Wall Corrugated Shipping Containers Using Fixed and Floating Text Platens" (*J. Testing and Evaluation* (1992):318-320) describes an experiment in which several different types of boxes were compared with respect to compression strength (lb). The treatment is the type of box and the response variable is compression strength (lb).

Type of Box

1	2	3	4
655.5	789.2	737.1	535.1
788.3	772.5	639.0	628.7
734.3	786.9	696.3	542.4
721.4	686.1	671.7	559.0
679.1	732.1	717.2	586.9
699.4	774.8	727.1	520.0

Follow these steps to test the hypothesis

1. Enter data. $H_0 : \mu_1 = \mu_2 = \mu_3 = \mu_4$

   Enter the data into four columns. In column 1, enter the data for Type1. Name column 1 as Type1. In column 2, enter the data for Type2. Name column 2 as Type2. In column 3, enter the data for Type3. Name column 3 as Type3. In column 4, enter the data for Type4. Name column 4 as Type4. The manner in which this data is entered is referred to as unstacked data.

2. Perform the ANOVA.

   Choose <u>S</u>tat><u>A</u>NOVA>Oneway (<u>U</u>nstacked). Place Type1, Type2, Type3 and Type4 in the Responses (in separate columns): text box.

3. Construct boxplots of the data.

   Choose **Gr<u>a</u>phs**. Place a check in the <u>B</u>oxplots of data checkbox. Choose <u>O</u>K. Choose <u>O</u>K.

The Minitab Output

```
One-Way Analysis of Variance

Analysis of Variance
Source      DF         SS        MS        F        P
Factor       3     127423     42474    25.10    0.000
Error       20      33847      1692
Total       23     161270
                                   Individual 95% CIs For Mean
                                   Based on Pooled StDev
Level        N       Mean     StDev   ----+---------+---------+---------+--
Type1        6     713.00     46.55                          (---*----)
Type2        6     756.93     40.34                              (----*--
Type3        6     698.07     37.20                       (---*-----)
Type4        6     561.98     39.89   (---*-----)

Pooled StDev =    41.14                560       640       720
```

Figure 14.1

The Minitab output (Figure 14.1) indicates the One-Way Analysis of Variance table. The output contains the ANOVA summary table with the sources of variation, associated degrees of freedom, mean square, F statistic and the associated p-value. The F statistic ($F = 25.10$) and associated p-value ($p = 0.000$) indicates that there is compelling evidence to reject the null hypothesis $H_0 : \mu_1 = \mu_2 = \mu_3 = \mu_4$ at the .05 level of significance. Observe that there is a substantial amount of overlap in the individual 95% confidence intervals for Type1, Type2 and Type3. This indicates that there are no significant differences between these types. On the other hand the compression strength for Type 4 is considerably smaller than Type1, Type2 and Type3. This indicates that there is a significant difference between Type4 and

195

the other types.

The Minitab Output

Boxplots of Type1 - Type4
(means are indicated by solid circles)

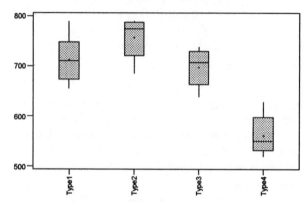

Figure 14.2

The Minitab output (Figure 14.2) indicates the comparative boxplot for the four types of boxes. Observe that the compression strengths for Type4 appears to be considerably smaller than the other types. This boxplot is consistent with the 95% individual confidence intervals examined in Figure 13.1.

The Problem - Storage Times

The accompanying data on calcium content of wheat is consistent with summary quantities that appeared in the paper "Mineral Contents of Cereal Grains as Affected by Storage and Insect Infestation" (*J. Stored Prod. Res.* (1992):147-151). Four different times were considered.

	Storage Period		
0 Months	1 Month	2 Months	4 Months
58.75	58.87	59.13	62.32
57.94	56.43	60.38	58.76
58.91	56.51	58.01	60.03
56.85	57.67	59.95	59.36
55.21	59.75	59.51	59.61
57.30	58.48	60.34	61.95

Follow these steps to test the hypothesis

$$H_0 : \mu_1 = \mu_2 = \mu_3 = \mu_4$$

1. Enter data.

   Enter all of the data into one column. In column 1, enter the data for 0 Months followed by the data for 1 Month, followed by the data for 2 Months, followed by the data for 4 Months. Name column 1 as Calcium. In column 2, enter the codes for each storage period. Enter six 0's followed by six 1's, followed by six 2's, followed by six 4's. Name column 2 as Period. The manner in which this data is entered is referred to as stacked data.

2. Perform the ANOVA.
   Choose <u>Stat</u>><u>A</u>NOVA><u>O</u>neway. Place Calcium in the Response: text box. Place Period in the <u>F</u>actor: text box. Choose <u>O</u>K.

The Minitab Output

One-Way Analysis of Variance

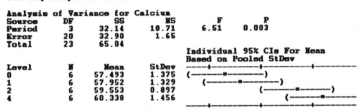

Figure 14.3

The Minitab output (Figure 14.3) indicates the One-Way Analysis of Variance table. The output contains the ANOVA summary table with the sources of variation, associated degrees of freedom, mean square, F statistic and the associated p-value. The F statistic ($F = 6.51$) and associated p-value ($p = 0.003$) indicates that there is compelling evidence to reject the null hypothesis $H_0 : \mu_1 = \mu_2 = \mu_3 = \mu_4$ at the .05 level of significance.

14.3   Multiple Comparisons

New Minitab Commands

1. <u>Stat</u>><u>A</u>NOVA><u>O</u>neway - Performs a one-way analysis of variance, with the dependent variable in one column, subscripts in another. In this section, you will perform a one-way analysis of variance on four samples, where all of the data is contained in one column, and the subscripts in another column.

   a. <u>Comparisons</u> - Provides confidence intervals for the differences between means, using four different methods: Tukey's, Fisher's, Dunnett's, and Hsu's MCB. Tukey and Fisher provide confidence intervals for all pairwise differences between level means. In this section, you will place a check in the <u>T</u>ukey's, family error rate: checkbox to perform a multiple comparison procedure.

When $H_0 : \mu_1 = \mu_2 = \mu_3 = \mu_4$ is rejected by the F test, we believe that there are differences among the k population means. The question is "Which pairs of means differ?". A multiple comparison procedure is a process for identifying differences among the k population means once the hypothesis of overall equality has been rejected. One of those methods is the Tukey-Kramer multiple comparison method.

The Tukey-Kramer multiple comparison method provides a confidence interval for all pairwise differences between means. The null hypothesis of no difference between two means may be rejected if and only if the confidence interval does not include zero.

The Problem - Peanut Butter
A study of the nutritional content of various types of peanut butter revealed the following total fat content (grams). The data is contained in the file a:pbutter.mtp.

Type of Peanut Butter

Creamy	Chunky	Other
16	16	16
15	16	12
16	16	12
16	17	18
16		11
16		12
16		
17		
16		
16		

Follow these steps to test the hypothesis

$$H_0 : \mu_1 = \mu_2 = \mu_3$$

1. Enter data.
   Enter all of the data into one column. In column 1, enter the data (amount of total fat) for Creamy followed by the data for Chunky followed by the data for Other types of peanut butter. Name column 1 as TotalFats. In column 2, enter the codes for each type of peanut butter. Enter Creamy, Chunky and Other as appropriate. Name column 2 as Type. The manner in which this data is entered is referred to as stacked data.

2. Perform the ANOVA.
   Choose **Stat**>**ANOVA**>**Oneway**. Place TotalFat in the Response: text box. Place Type in the **F**actor: text box.

3. Make comparisons.
   Choose **Comparisons**. Place a check in the **T**ukey's, family error rate: checkbox. Choose the default value of 0.05. Choose **OK**. Choose **OK**.

The Minitab Output

```
One-Way Analysis of Variance
Analysis of Variance for TotalFat
Source      DF         SS         MS           F        P
Type         2      27.95      13.98        5.62    0.013
Error       17      42.25       2.49
Total       19      70.20
                                        Individual 95% CIs For Mean
                                        Based on Pooled StDev
Level        N       Mean      StDev    ----+---------+---------+---------+-----
Chunky       4     16.250      0.500                      (----------•------)
Creamy      10     16.000      0.471                         (------•-------)
Other        6     13.500      2.811    (--------•---------)
                                        ----+---------+---------+---------+-----
Pooled StDev =       1.576             12.0      14.4      16.0      1
```

Figure 14.4

The Minitab output (Figure 14.4) indicates the One-Way Analysis of Variance table. The output contains the ANOVA summary table with the sources of variation, associated degrees of freedom, mean square, F statistic and the associated p-value. The F statistic ($F = 5.62$) and associated p-value ($p = 0.013$) indicates that there is compelling evidence to reject the null hypothesis $H_0 : \mu_1 = \mu_2 = \mu_3$ at the .05 level of significance.

Once the null hypothesis has been rejected (Figure 13.4), the question is to determine which pairs of means are significantly different.

The Minitab Output

```
Tukey's pairwise comparisons

       Family error rate = 0.0500
   Individual error rate = 0.0201

Critical value = 3.63

Intervals for (column level mean) - (row level mean)

             Chunky      Creamy

Creamy      -2.144
             2.644

Other        0.138       0.410
             5.362       4.590
```

Figure 14.5

The Minitab output (Figure 13.5) indicates that only the intervals $\mu_{Chunky} - \mu_{Other}$ and $\mu_{Creamy} - \mu_{Other}$ do not contain zero. We therefore judge μ_{Other} to differ significantly from both μ_{Chunky} and μ_{Creamy}, and these are the only two significant differences. The total fat (grams) content for the Other types (reduced fat, smart choice) is significantly less than Creamy and Chunky types of peanut butter.

Exercises

14.1 The accompanying data resulted from a flammability study in which specimens of five different fabrics were tested to determine burn times.

Fabric

1	2	3	4	5
17.8	13.2	11.8	16.5	13.9
16.2	10.4	11.0	15.3	10.8
15.9	11.3	9.2	14.1	12.8
15.5		10.0	15.0	11.7
			13.9	

a. Perform the appropriate hypothesis test using a .05 level of significance.

b. If multiple comparisons are appropriate, use the Tukey-Kramer multiple comparison procedure and state the results; if multiple comparisons are not necessary, explain why not.

14.2 The accompanying data resulted from a study to determine the weight of water absorped by four different brands of paper towels.

Paper Towel

A	B	C	D
114	122	108	122
106	118	116	112
112	124	112	114
110	116	110	114
114	114	118	116
116	120	114	118

a. Perform the appropriate hypothesis test using a .05 level of significance.

b. If multiple comparisons are appropriate, use the Tukey-Kramer multiple comparison procedure and state the results; if multiple comparisons are not necessary, explain why not.

14.3 The concentration of a catalyst used in the production of a certain type of plastic is thought to affect its strength. An experiment was conducted to examine the effects of three different concentrations of the catalyst resulting in the following data.

Strength

2%	4%	6%
7.9	8.6	8.1
6.5	8.1	8.8
5.8	7.0	9.2
8.3	9.5	10.1
6.1	8.4	8.9

a. Perform the appropriate hypothesis test using a .05 level of significance.

b. If multiple comparisons are appropriate, use the Tukey-Kramer multiple comparison procedure and state the results; if multiple comparisons are not necessary, explain why not.

14.4 The fog index is a measure of reading difficulty based on the average number of words per sentence and the percentage of words with three or more syllables. High values of the fog index are associated with difficult reading levels. Independent random samples of six advertisements were taken from three different magazines and fog indices were computed to obtain the data given in the accompanying table (*"J. of Ad. Research* (1981):45-50).

Scientific American	Fortune	New Yorker
15.75	12.63	9.27
11.55	11.46	8.28
11.16	10.77	8.15
9.92	9.93	6.37
9.23	9.87	6.37
8.20	9.42	5.66

a. Perform the appropriate hypothesis test using a .01 level of significance.

b. If multiple comparisons are appropriate, use the Tukey-Kramer multiple comparison procedure and state the results; if multiple comparisons are not necessary, explain why not.

14.4   Randomized Block Experiments

New Minitab Commands

1. <u>S</u>tat><u>A</u>NOVA><u>T</u>woway - Performs a two-way analysis of variance for balanced data. Each cell must contain an equal number of observations. You can not specify whether the effects are fixed or random with TWOWAY. As a result, TWOWAY does not produce F and p-values as Minitab would have to guess at the type of effects you have. In this section, you will use this command to perform a two-way analysis of variance where four treatments and five blocks are present.

In Chapter 10, we saw that when two treatments are to be compared, a paired experiment is often more effective than one involving two independent samples. For example, in a study of the effects of different diets on weight loss, subjects are often paired (blocked or grouped) into different initial weight categories. Within each initial weight category, the subjects are alike as possible. Then within each initial weight category, one subject is randomly selected for diet 1, a second subject is randomly selected to receive diet 2, and so on. These homogeneous groups (initial weight categories) are called blocks, and the random allocation of treatments (diets) within each block as described gives a randomized block experiment.

Let experimental units (individuals or objects to which the treatments are applied) be separated into groups consisting of k units in such a way that the units within each group are as similar as possible. Each unit in a group receives a different treatment. The groups are often called blocks, and the experimental design is referred to as a randomized block design.

Chapter   14

The Problem - Energy Efficiency
A large developer carried out a study to compare electricity usage for four different
residential air-conditioning systems being considered for tract homes. Each system
was installed in five homes, and the resulting electricity usage in (KWh) was monitored for a one-month period. Homes selected for the experiment were grouped
into five blocks consisting of four homes each so that the four homes within any
given block were as similar as possible.

	Block				
Treatment	1	2	3	4	5
1	116	118	97	101	115
2	171	131	105	107	129
3	138	131	115	93	110
4	144	141	115	93	99

Follow these steps to test the hypothesis of interest that the mean value does not
depend on which treatment is applied:

$$H_0 : \mu_1 = \mu_2 = \mu_3 = \mu_4$$

1. Enter data.
   Enter all of the data into one column. In column 1, enter the data for Block 1,
   followed by the data for Block 2, followed by the data for Block 3, followed
   by the data for Block 4, followed by the data for Block 5. Name column 1 as
   KWh.

2. Enter codes for blocks.
   Choose **Calc**>**Make Patterned Data**>**Simple Set of Numbers**. Place Blocks
   in the Store patterned data in: text box. Place 1 in the From first value: text
   box. Place 5 in the To last value: text box. Place 1 in the In steps of: text
   box. Place 4 in the List each value: ____ times text box. Place 1 in the List the
   Whole sequence ____ times text box. Choose **OK**.

3. Enter codes for treatments.
   Choose **Calc**>**Make Patterned Data**>**Simple Set of Numbers**. Place Treatment in the Store patterned data in: text box. Place 1 in the From first value:
   text box. Place 4 in the To last value: text box. Place 1 in the In steps of: text
   box. Place 1 in the List each value: ____ times text box. Place 5 in the List the
   Whole sequence ____ times text box. Choose **OK**.

4. Perform the ANOVA.
   Choose **Stat**>**ANOVA**>**Twoway**. Place KWh in the Response: text box. Place
   Treatment in the Row factor: text box. Place Blocks in the Column factor: text
   box. Choose **OK**.

The Minitab Output

Two-way Analysis of Variance

```
Analysis of Variance for KWh
Source        DF        SS        MS
Treatmen       3       930       310
Blocks         4      4960      1240
Error         12      1705       142
Total         19      7595
```

Figure 14.6

The Minitab output (Figure 14.6) indicates the Two-Way Analysis of Variance table. The output contains the ANOVA summary table with the sources of variation, associated degrees of freedom, and mean square. (You can not specify whether the effects are fixed or random with TWOWAY. As a result, TWOWAY does not produce F and p-values as Minitab would have to guess at the type of effects you have. Use Balanced ANOVA to perform a two-way analysis of variance, specify fixed or random effects, and display the F and p-values when you have balanced data.). Since Minitab does not produce the F statistic, the value may be calculated by dividing the $MS_{Treatment}$ by MS_{Error} ($F = \frac{MS_{Treatment}}{MS_{Error}} = \frac{310}{142} = 2.18$). The F statistic ($F = 2.18$) is not significant at the .05 level of significance when compared to the critical value ($F_{12,3,\alpha=.05} = 3.49$). Mean electricity usage does not seem to depend on which of the four air-conditioning systems is used.

Exercises

14.5 The paper "Measuring Treatment Effects Through Comparisons Along Plot Boundaries" (Forest Sci. (1980):704-709) reported the results of a randomized block experiment. Five different sources of pine seed were used in each of the four blocks. The accompanying table gives data on plant height (m).

		Block		
Source	1	2	3	4
1	7.1	5.8	7.2	6.9
2	6.2	5.3	7.7	4.7
3	7.9	5.4	8.6	6.2
4	9.0	5.9	5.7	7.3
5	7.0	6.3	4.4	6.1

a. Perform the appropriate hypothesis test using a .05 level of significance.

14.6 An evaluation of the ability of fabric to repel water is performed. The main objective is to determine which of three different types of fabric, lableled fabric A, B, and C has the best ability to repel water. The fabrics are treated with different concentrations of a chemical, lableled 5% and 10%.

	Fabric		
Concentration	A	B	C
5%	68.0	72.9	74.2
10%	68.5	70.8	68.4

a. Perform the appropriate hypothesis test using a .05 level of significance.

14.7 An evaluation of the ability of plastic tubing to withstand internal pressure is performed. The amount of internal pressure necessary to ruputure the tubing is of interest. Three different types of plastic tubing, labeled tubing A, B, and C, were produced using two different processes, labeled method D and E.

	Tubing		
Method	A	B	C
D	70.1	79.3	81.1
E	75.8	83.0	83.7

a. Perform the appropriate hypothesis test using a .05 level of significance.

14.5   The Two-Factor ANOVA

New Minitab Commands

1. **Stat>Tables>Cross Tabulation** - Prints one-way or multi-way contingency tables and tables of statistics for associated variables, and displays the output in an easy-to-read format in the Session window. In this section, you will use this command to obtain a table of means for two factors (variables) in a two factor analysis of variance.

2. **Stat>ANOVA>Balanced ANOVA** - Performs univariate and multivariate analysis of variance. Factors may be crossed or nested, fixed or random. Nesting must be balanced and the subscripts used to indicate levels of B within each level of A must be the same. For a two-way analysis of variance data must be balanced (all cells have the same number of observations).

3. **Stat>ANOVA>Interactions Plot** - Draws a single interaction plot if 2 factors are entered. In this section, you will use this command to construct an interaction plot for two factors in a two-factor analysis of variance.

An investigator will often be interested in assessing the effects of two different factors on a response variable. For example, an agricultural scientist in determining how yield of tomatoes is affected by choice of variety planted (a categorical factor, with each category corresponding to a different variety, say variety 1, variety 2 and variety 3) and planting density (a quantitative factor, with a level corresponding to each planting density being considered, say 10, 20, 30 and 40 thousand plants per hectare).

Let's call the two factors under study factor A and factor B. Even when a factor is categorical, it simplifies terminology to refer to the categories as levels. Thus, the categorical factor variety may have a number of levels, one for each variety. The number of levels of factor A is denoted by k, and l denotes the number of levels of factor B. Each cell in the table corresponds to a particular level of factor A in combination with a particular level of factor B. Because there are l cells in each row and k rows, there are kl cells in the table. The kl different combinations of factor A and factor B levels are often referred to as treatments. In this tomato example, there are three tomato varieties and four different planting densities under consideration, providing $3 \times 4 = 12$ treatments.

An experimenter frequently designs the experiment to have, m, the same number of observations on each treatment.

The Problem - Tomato Yield

An experiment was carried out to assess the effects of tomato variety (factor A, with $k = 3$ levels) and planting density (factor B, with $l = 4$ levels - 10, 20, 30 and 40 thousand plants per hectare) on yield. Each of the $kl = 12$ treatments was used on $m = 3$ plots, resulting in the data set (adapted from "Effects of Plant Density on Tomato Yields in Western Nigeria", *Exper. Ag.* (1976):43-47).

			Density	
Variety	10	20	30	40
	7.9	11.2	12.1	9.1
1	9.2	12.8	12.6	10.8
	10.5	13.3	14.0	12.5
	8.1	11.5	13.7	11.3
2	8.6	12.7	14.4	12.5
	10.1	13.7	15.4	14.5
	15.3	16.6	18.0	17.2
3	16.1	18.5	20.8	18.4
	17.5	19.2	21.0	18.9

Sample average yields for each treatment, each level of factor A, and each level of

factor B are important summary quantities.

Follow these steps to obtain a table of means.

1. Enter data.

   Enter all of the data into one column. In column 1, enter the data for planting density 10, followed by the data for planting density 20, followed by the data for planting density 30, followed by the data for planting density 40. Name column 1 as Yield.

2. Enter codes for densities.

   Choose **Calc>Make Patterned Data>Simple Set of Numbers.** Place Density in the Store patterned data in: text box. Place 10 in the From first value: text box. Place 40 in the To last value: text box. Place 10 in the In steps of: text box. Place 9 in the List each value: ____ times text box. Place 1 in the List the Whole sequence ____ times text box. Choose **OK.** In this step, you have created the subscripts corresponding to the density for each observation.

3. Enter codes for varieties.

   Choose **Calc>Make Patterned Data>Simple Set of Numbers.** Place Variety in the Store patterned data in: text box. Place 1 in the From first value: text box. Place 3 in the To last value: text box. Place 1 in the In steps of: text box. Place 3 in the List each value: ____ times text box. Place 4 in the List the Whole sequence ____ times text box. Choose **OK.** In this step, you have created the subscripts corresponding to the variety for each observation.

4. Obtain the table of means.

Choose **Stat**>**Tables**>**Cross Tabulation**. Place Variety in the Classification variables: text box. Place Density in the Classification variables: text box. Choose **Summaries**. Place Yield in the Associated variables: text box. Place a check in the Display **Means** checkbox. Choose **OK**. Choose **OK**.

The Minitab Output

Tabulated Statistics

Rows: Variety   Columns: Density

	10	20	30	40	All
1	9.200	12.433	12.900	10.800	11.333
2	8.933	12.633	14.500	12.767	12.208
3	16.300	18.100	19.933	18.167	18.125
All	11.478	14.389	15.778	13.911	13.889

Cell Contents --
                Yield:Mean

Figure 14.7

The Minitab output (Figure 14.7) indicates the sample mean for each cell, each variety (row) and each planting density (column) .

Interaction

An important aspect of two-factor studies involves assessing how simultaneous changes in the levels of both factors affect the response. When there are more than two levels of either factor, a graph of true average responses provides insight into how changes in the level of one factor depend on the level of the other factor. An interaction of the factors is suggested when the change in the true average response occurs when the level of one factor changes depend on the level of the other factor. Graphically, there is no interaction between the factors when the lines are parallel. Conversely, an interaction between two factors is suggested when the lines are not parallel.

Hypotheses for a Two-Factor ANOVA

The hypothesis of primary interest in a two-factor ANOVA involves interaction.

H_0 : There is no interaction between factors.

H_a :There is an interaction between factors.

The following two hypotheses should be tested only if H_0 :There is no interaction between factors is not rejected.

H_0 :There are no factor A main effects.

H_a : at least two of the μ's for factor A are different.

H_0 :There are no factor B main effects

H_a : at least two of the μ's for factor B are different

The Problem - Tomato Yield
Follow these steps to test the hypothesis that there is no interaction between tomato varieties and plant density for the tomato data.

1. Perform the ANOVA.

   Choose **Stat**>**ANOVA**>**Balanced ANOVA.** Place Yield in the Responses: text box. Place Density Variety Density*Variety in the Model: text box. (The Density*Variety represents the interaction term.) Choose **OK**.

The Minitab Output

Analysis of Variance (Balanced Designs)

```
Factor      Type Levels Values
Density     fixed     4     10    20    30    40
Variety     fixed     3      1     2     3

Analysis of Variance for Yield

Source            DF        SS        MS       F      P
Density            3    86.687    28.896   18.23  0.000
Variety            2   327.597   163.799  103.34  0.000
Density*Variety    6     8.032     1.339    0.84  0.548
Error             24    38.040     1.585
Total             35   460.356
```

Figure 14.8

The Minitab output (Figure 14.8) indicates the Analysis of Variance (Balanced Designs) table. The output contains the ANOVA summary table with the sources of variation, associated degrees of freedom, mean square, F statistics and associated p-values. The F statistic for interaction is 0.84 with an associated p-value of 0.548. The F statistic ($F = 0.84$) is not significant at the .05 level of significance when compared to the critical value ($F_{24,6,\alpha=.10} = 2.51$), so it is appropriate to examine the analysis for the presence of main effects.

The F statistic for Density is 18.23 with an associated p-value of 0.000. Therefore, H_0 : There are no factor B (Density) main effects is rejected, and we conclude that the true average yield does depend on which planting density is used.

The F statistic for Variety is 103.34 with an associated p-value of 0.000. Therefore, H_0 : There are no factor A (Variety) main effects is rejected, and we conclude that the true average yield does depend on which plant variety is used.

After the null hypothesis of no factor A (or factor B) main effects has been rejected, significant differences in factor A (factor B) levels can be identified by using the Tukey-Kramer multiple comparison method described earlier.

Recall that the F statistic for interaction was 0.84 with an associated p-value of 0.548. Consequently, the null hypothesis H_0 :There is no interaction between factors was not rejected.

Chapter 14

Let's see what happens when an interaction plot is constructed.
Follow these steps to construct an interaction plot.
1. Construct the plot.
   Choose **Stat>ANOVA>Interactions Plot.** Place Variety in the Factors: text box. Place Density in the Factors: text box. Place Yield in the Raw response data in: text box. Choose **OK.**

The Minitab Output

Interaction Plot - Means for Yield

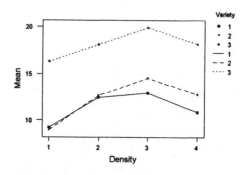

Figure 14.9

The Minitab output (Figure 14.9) indicates the interaction plot for the data with Density as the horizontal axis. Observe that the lines are reasonably parallel, implying no significant interaction effects. This conclusion is consistent with the ANOVA table.

Exercises

14.8 The effect of two factors, humidity and coating, on the lifetime of a particular product was examined.

Coating	Humidity		
	Low (20%)	Medium (50%)	High (85%)
A	2.51	2.50	2.18
	2.64	2.61	2.23
	2.69	2.65	2.12
B	2.97	2.06	3.71
	3.01	2.12	1.59
	3.10	2.18	2.64

a. Use a significance level of .01 to test the null hypothesis of no interaction between coating and humidity.
b. Construct an interaction plot with humidity on the horizontal axis.

14.9 The effect of three different soil type and two different varieties of wheat on wheat height were examined in a study.

Variety	Soil Type			
	A	B	C	D
	35.0	38.8	31.1	34.0
E	31.7	45.8	26.1	40.1
	39.8	41.7	29.2	37.5
	40.6	47.4	45.4	36.2
F	38.3	49.6	42.0	34.3
	45.0	43.2	50.6	30.3

a. Use a significance level of .01 to test the null hypothesis of no interaction between soil type and variety.

b. Construct an interaction plot with soil type on the horizontal axis.